Study Guide for
STATISTICS FOR BUSINESS AND FINANCIAL ECONOMICS

Second Edition

Study Guide for
STATISTICS FOR BUSINESS AND FINANCIAL ECONOMICS

Second Edition

Cheng F. Lee, John C. Lee & Alice C. Lee

Ronald L. Moy
St. John's University, New York

World Scientific
Singapore • New Jersey • London • Hong Kong

Published by

World Scientific Publishing Co. Pte. Ltd.

P O Box 128, Farrer Road, Singapore 912805

USA office: Suite 1B, 1060 Main Street, River Edge, NJ 07661

UK office: 57 Shelton Street, Covent Garden, London WC2H 9HE

STUDY GUIDE FOR STATISTICS FOR BUSINESS AND FINANCIAL ECONOMICS (Second Edition)

ISBN 981-02-3831-2 (pbk)

Printed in Singapore by Uto-Print

PREFACE

Nearly all students in business and economics regard a course in statistics as a "necessary evil" required for graduation. I have known of students who postponed taking the required course in statistics until the final semester of their senior year in hope that they will receive sympathy from the instructor simply because they are graduating.

While many students suffer from this anxiety, I believe that much of their fear is unwarranted. Aside from providing important skills used in business decision making, the use of statistics can actually be fun. Sports fans, for instance, constantly deal with statistics, from the average yards a running back gains every time he carries the ball, to the percentage of first serves that are good for a tennis player, to the batting average of a baseball player. Those of you who are not sports fans also deal with statistics in your everyday life. For example, the television ratings provided by A. C. Nielsen and Company, the polls that predict the winners of elections, and the probability of snow from a weather forecast all represent statistics. The person who has a basic understanding of statistics will have a great advantage throughout life. You can be sure that when the time comes for contract negotiations, actor Tom Hanks and baseball player Mark McGuire know statistics.

To reflect the many uses of statistics, I have used examples and problems from all walks of life, including sports, politics, and entertainment as well as business and economics.

The organization of this study guide closely parallels that of Cheng F. Lee's *Statistics for Business and Financial Economics*, providing a comprehensive treatment of every chapter. As an educator, I realize that there are many ways to present a topic. What works for one student may not work for another. To give students the best opportunity to understand the material, I try to present the text material in a slightly different although complementary way. For each chapter, the study guide provides:

- ***Chapter Intuition***. Each chapter begins with an intuitive verbal explanation of the chapter's central message. Essentially, I try to provide some sense of why this chapter is important and where it is headed.

- ***Chapter Review***. Rather than just giving a simple outline of the chapter, all the key concepts in the chapter are presented in a simple, easy-to-follow narrative.

- ***Useful Formulas***. When appropriate, a list of useful formulas from the chapter is presented so that you will not need to search through the text to find formulas necessary for solving the problems.

- ***Example Problems and Solutions***. Here, sample problems similar to the problems in the text are presented, along with a step-by-step solution. To provide you with a guide to solving the problems, each example states the topic that the problem presents.

- ***Supplementary Exercises***. Once you have studied the example problems, you can begin to put your skills to work by solving problems. A variety of exercise types is offered to accommodate various learning styles.

I hope that this study guide will make your study of statistics easier, and more interesting, and perhaps even fun. Good luck.

I would like to thank C. F. Lee for giving me the opportunity to work on this project.

R. L. M.

CONTENTS

CHAPTER 1

INTRODUCTION

Chapter Intuition

Throughout our day-to-day activities, we are constantly bombarded with numbers. Whether we are checking out the results from last night's baseball game or reading the financial pages in the newspaper, we must deal with numbers. One problem with simply looking at a large amount of numbers is that it can be difficult to interpret what they mean. One way to deal with these numbers is to find a method for organizing them so they can be easily understood.

Statistics is an approach that allows us to systematically organize, present and analyze data. There are two basic uses of statistics: to compare or describe data and to make inferences about uncertain events. When statistics is used to describe or compare, we call it ***descriptive statistics***. Most of us have seen a wide variety of descriptive statistics. Sports enthusiasts are especially bombarded with descriptive statistics. Average points per game for a basketball player, average yards rushing for a football player, and batting averages in baseball all represent descriptive statistics. Those of you who are not interested in sports also deal frequently with descriptive statistics such as your grade point average, and the average salary a graduating senior receives. Descriptive statistics is important because it allows us to summarize a series of numbers such as the grades you received in all your courses into a single number, your grade point average. Also, descriptive statistics enables us to make comparisons. For example, if you received a score of 80 on your statistics midterm, it would be difficult for you to know how well you did without knowing how other students in the class performed. By looking at the class average, you can compare your score to the performance of the rest of the class.

When we use statistics to make guesses about uncertain things, we are using an approach known as ***inferential statistics***, because it allows us to infer something that is not yet known. For example, polls that predict the outcome of elections use inferential statistics. Inferential statistics is an approach to making educated guesses about what an entire population (for example, all voters) would do based on a smaller sample (for example, a selection of 1,000 voters). Inferential statistics is especially important when it is too costly or time consuming to survey the entire population.

Chapter Review

1. ***Statistics*** is a course of study devoted to the collection, organization and analysis of data.
2. The entire group we are interested in studying is called the ***population***. A ***sample*** is a subset of the population.
3. ***Descriptive statistics*** allows us to compare different things. For example, we can use a descriptive statistic like the mean (or average) to compare the height of the Boston Celtics to that of the Los Angeles Lakers. Descriptive statistics can be used for comparisons between different groups as in the previous example or it can be used to measure performance such as the return of AT&T's stock, or the average yards Eric Dickerson gains every time he carries the ball, or the average SAT score for students entering Stanford University or the average performance of a statistics class on the midterm exam. Whether we realize it or not, all of us have dealt with descriptive statistics in our lives. The first part of the text is devoted to descriptive statistics.
4. In many cases, we are interested in going beyond simply describing a set of data. ***Inferential statistics*** allows us to draw conclusions about an entire population using only a smaller subset of the population. For example, inferential statistics lets us draw conclusions about how all voters will vote in the presidential election by looking at how only a few will vote. The benefit of using inferential statistics is that it enables us to draw reasonably accurate conclusions inexpensively and efficiently. Imagine the time, money and energy that would be necessary to survey all registered voters in the U.S. about their preferences for the various presidential candidates. The second half of the book is devoted to inferential statistics.
5. ***Deductive statistics*** draws conclusions about specific cases using general information.
6. ***Inductive statistics*** draws general conclusions based on specific information.

Example Problems

Example 1 **Descriptive vs. Inferential Statistics**

Explain whether each of the following is a descriptive or inferential statistic:

a. The average earnings per share for AT&T over the last 5 years.

b. The number of people who will vote for the Democratic candidate for senator in the upcoming election in California using a sample of 200 potential voters.

c. The number of people who would favor a constitutional amendment requiring Congress to balance the budget, based on a survey of registered voters.

d. The average number of yards a rookie running back is expected to gain, based on a sample of rookie running backs.

Solution:

a. Because averages are used to summarize or describe past data, they are descriptive statistics. So the average EPS is a descriptive statistic.

b. When we use a small subset of data to guess or infer the behavior of an entire population, we are using inferential statistics.

c. Inferential statistic

d. Inferential statistic

Example 2 **Using Descriptive Statistics**

Briefly explain how the agent for Greg Maddox of the Atlanta Braves could have used statistics to negotiate a contract similar to that of the Los Angeles Dodgers Kevin Brown in 1998.

Solution: For Maddox's agent to negotiate a contract similar to the one Kevin Brown signed in early 1998, he could have shown that the pitching performance of the two players was similar. To do this, he could have used statistics such as the earned run average of each pitcher (the average number of runs given up per 9 innings pitched), the winning percentages of each player, average strikeouts per year, and so on.

Example 3 **Using Descriptive Statistics**

Suppose you are graduating this semester. What statistics might be useful in helping you negotiate a fair salary?

Solution: A good starting point for your negotiations would be the average starting salary for other students in your major. However, other statistics could be useful too. First, if you have a very good grade point average, you could argue that you deserve an above average starting salary simply because you are well above average. Another useful statistic would be the starting salaries for other fields of study in which you have taken courses. For example, you are graduating with a major in economics and a minor in computer science. If the starting salary for economics majors is $35,000 and the starting salary for computer science majors is $42,000, you may be able to argue that your computer science background entitles you to be paid more in line with computer science majors rather than being paid as an economics major.

Supplementary Exercises

Multiple Choice

1. Statistics is a
   a. method for organizing and analyzing data.
   b. method for describing data.
   c. method for making educated guesses based on limited data.
   d. method for summarizing data.
   e. all of the above.

2. Descriptive statistics can be used to
   a. compare different sets of data.
   b. compare one observation to a set of data.
   c. make inferences about unknown events.
   d. make guesses about a population using a sample.
   e. both a and b.

3. The average points per game for a basketball player represents
   a. a descriptive statistic.
   b. an inferential statistic.
   c. a deductive statistic.
   d. an inductive statistic.
   e. a sample.

4. Using the Nielsen television ratings to estimate the number of television viewers represents
   a. descriptive statistics.
   b. inferential statistics.
   c. deductive statistics.
   d. inductive statistics.
   e. a population.

5. Using the Gallup election poll to predict the outcome of an election represents
   a. descriptive statistics.
   b. inferential statistics.
   c. deductive statistics.
   d. inductive statistics.
   e. a population.

6. Your grade point average is
   a. descriptive statistics.
   b. inferential statistics.
   c. deductive statistics.
   d. inductive statistics.
   e. a sample.

7. When you are interested in comparing data you would use
   a. descriptive statistics.
   b. inferential statistics.
   c. deductive statistics.
   d. inductive statistics.
   e. a census.

8. Business people use
   a. descriptive statistics.
   b. inferential statistics.
   c. deductive statistics.
   d. inductive statistics.
   e. all of the above.

9. If you were interested in predicting the outcome of the presidential election, you would use
   a. descriptive statistics.
   b. inferential statistics.
   c. deductive statistics.
   d. inductive statistics.
   e. a census.

10. If a business is interested in predicting which flavor mouthwash will be favored by all consumers based on a sample of 500 people they should use
    a. descriptive statistics.
    b. inferential statistics.
    c. deductive statistics.
    d. inductive statistics.
    e. a census.

True/False (If false, explain why)

1. When we use the Nielsen television ratings are used to estimate the number of television viewers, we are using descriptive statistics.
2. The average salary for a graduating senior majoring in finance is a descriptive statistic.
3. Descriptive statistics can be useful in contract negotiations.
4. Inferential statistics can be used to compare two sets of data.
5. Descriptive statistics is used to make educated guesses about unknown events.
6. Statistics can be useful in organizing and presenting data.
7. Deductive statistics draws general conclusions based on specific information.
8. Inductive statistics draws specific conclusions based on general information.
9. Descriptive statistics is frequently used in court cases to make comparisons.
10. The true average salary a graduating senior receives, based on a sample of 100 graduating seniors, is a descriptive statistic.

Questions and Problems

1. List three descriptive statistics commonly encountered in college.
2. List four descriptive statistics in baseball.
3. List four descriptive statistics commonly used in business.
4. Name two well-known sources of inferential statistics.

5. If you were interested in knowing what percentage of the 12 students in your art class will be attending the Picasso exhibit, would it be better to use a sample or a census? Why?
6. If you were interested in knowing what percentage of the 2,000 students in your school will be attending the Picasso exhibit, would it be better to use a sample or a census? Why?

Answers to Supplementary Exercises

Multiple Choice

1. e
2. e
3. a
4. b
5. b
6. a
7. a
8. e
9. b
10. b

True/False

1. False. Using the Nielsen television ratings to predict the number of television viewers represents inferential statistics.
2. True
3. True
4. False. Descriptive statistics is used to compare two sets of data.
5. False. Inferential statistics is used to make educated guesses.
6. True
7. False. Inductive statistics draws general conclusions based on specific information.
8. False. Deductive statistics draws specific conclusions based on general information.
9. True
10. False. Inferential statistics.

Questions and Problems

1. There are many descriptive statistics you may encounter in college including: grade point average, mean SAT score of all freshmen, average high school class rank, average starting salary for graduating seniors, percentage of female students, percentage of minorities, and average GMAT score.
2. Again, there are many descriptive statistics including: batting average, slugging percentage, earned run average, average strikeouts per nine innings pitched, and stolen base percentage.
3. In the business world, descriptive statistics might include: average advertising dollars, average sales per month, average revenues per month, average salary of employees, and average bonus per employee.
4. Probably the two most widely recognized sources of inferential statistics are the Nielsen television ratings, and the Gallup polls on voting preferences.
5. Because the size of the population is small, it is probably easier to use a census and simply ask all 12 people whether they are going to the exhibit or not.
6. Because the size of the population is relatively large, it may be better to use a sample of students and use inferential statistics to determine the percentage of all students who will be attending the exhibit.

CHAPTER 2

DATA COLLECTION AND PRESENTATION

Chapter Intuition

The say a "picture is worth a thousand words," and no where is this more true than when you are dealing with statistics. Once the data have been collected, they need to be organized and presented so they can be easily understood. For example, when your professor grades the midterm exam, she will have one score for each student. If the class is extremely large, it will be difficult for her to draw any conclusions about the class's performance by simply looking at a bunch of scores. However, by organizing the data into tables or graphs, she can much more easily determine how the class performed and how to assign grades. This chapter discusses how to collect and present data in an orderly manner so that it can be easily analyzed.

Chapter Review

1. Data can come from either a ***primary source***, which means it is collected specifically for the study, or from a ***secondary source***, which means that the data were originally collected for some other purpose.
2. Before collecting the data, the researcher must decide whether to take a ***sample*** or a ***census***. In a census, all members of the population of interest are surveyed. A census is most practical when the population is small. For example, when trying to decide which movie to see, a family of five would find little difficulty in surveying all five members. However, when the population of interest is very large, it may be infeasible to survey every member. In this case, it may be more desirable to take a sample of the entire population. In a sample, a subset of the population is used to evaluate the views of the entire population. For example, if we are interested in the views of all the men and women in the U.S. concerning a national health care policy, it would not be feasible to survey every single man and woman. Instead, we choose to examine the views of a

smaller number of men and women and attempt to make inferences about the entire population based on this sample.

3. There are two types of errors that are associated with primary and secondary data. ***Random error*** is the difference between the value obtained by taking a random sample and the value obtained by taking a census. ***Systematic error*** results when there are problems in measurement.

4. One way to organize data is to place them into groups, or classes, and then to present the data using tables.

5. Charts and graphs are another way for presenting data.

   a. A ***pie chart*** shows how the "whole," the pie, is divided into different pieces, the pieces of the pie. For example, a company could see how its advertising dollars are divided among television, radio, and newspapers by looking at a chart like the one below:

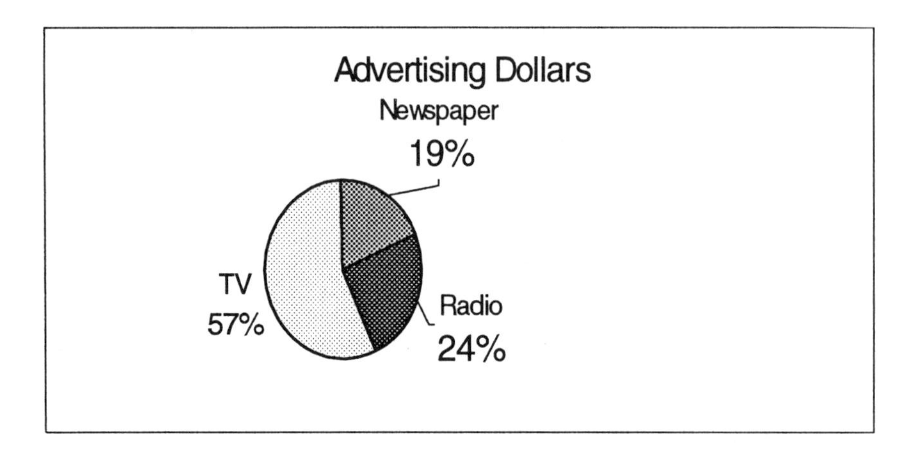

b. A ***bar chart*** can also be used to display data. The company in part (a), might use a bar graph to see how its advertising dollars are spent on television, radio, and newspapers. Each bar represents the amount spent on advertising for each medium.

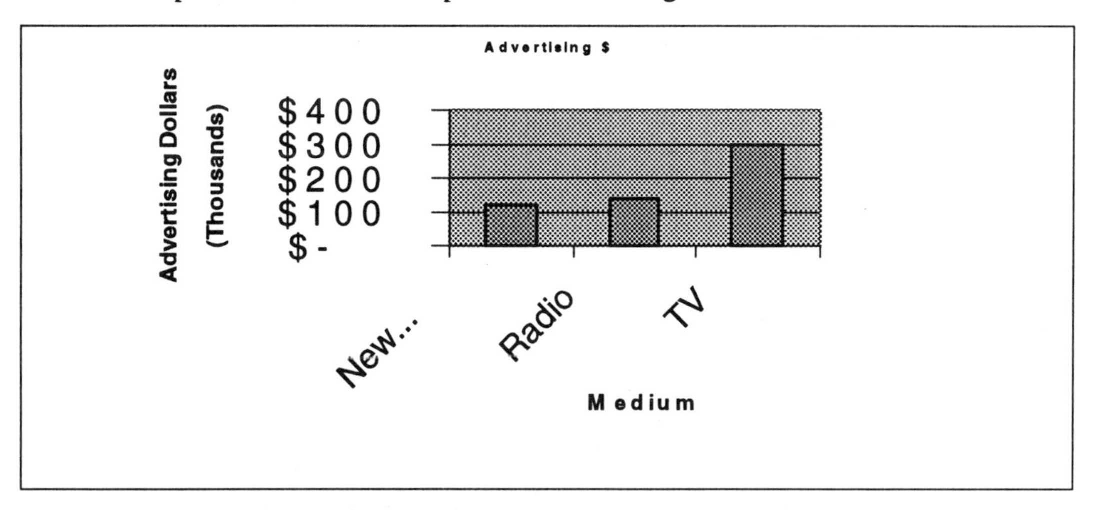

c. A ***line chart*** can be used to show the relationship between two different variables or how one variable changes over time. For example, the company in part (a) might be interested in seeing how its advertising expenditures have changed over time.

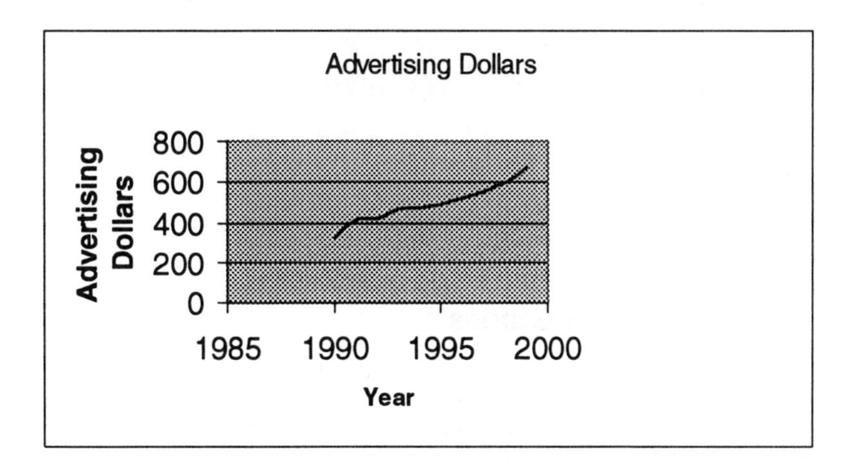

d. A ***time series graph*** is a line graph in which the variable on the X-axis represents time, such as the year or month of the data. The line graph shown for part (c) is also a time series graph.

Example Problems

Example 1 **Presenting Data in Graphs**

An economics professor gives the following grades to her class:

# of students	Grade
5	A
12	B
30	C
15	D
3	F

a. Use a bar chart to show the distribution of grades.

b. Use a pie chart to show the distribution of grades.

c. Which of these graphs do you think is best for presenting the distribution of grades? Why?

Solution:

a.

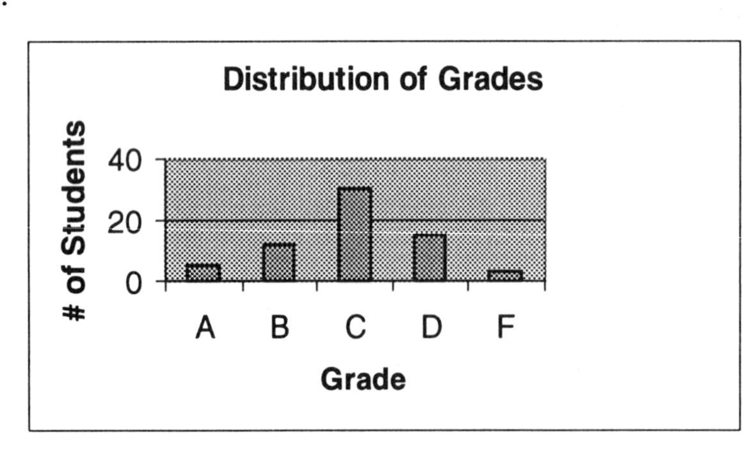

b.

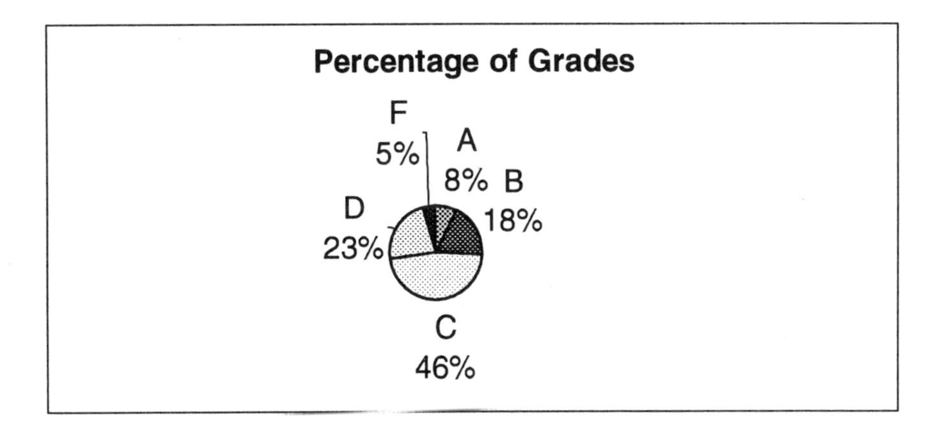

c. Both graphs do a good job of presenting the data. However, each graph stresses different aspects of the data. The bar chart gives us a good idea of where our grade stands with respect to others. For example, if we received a grade of C, we can see from the bar chart that most students received a grade of C and we can also easily see the number of students who had received higher and lower grades. On the other hand, the pie chart is better for showing us how the teacher divided the grades among the 65 students. Pie charts are best used when we are interested in seeing how the whole is divided among smaller subgroups.

Example 2 **Primary vs. Secondary Sources of Data**

Which of the following are from primary sources and which are from secondary sources:

a. The Dow Jones Industrial Average taken from the *Wall Street Journal.*

b. Responses to a survey on how chief financial officers will respond to a change in the accounting rules.

c. Johnson & Johnson's earnings as given in its annual report.

Solution:
a. secondary source
b. primary source
c. secondary source

Supplementary Exercises

Multiple Choice

1. Data from primary source
   a. are collected for other purposes than the current study.
   b. can be obtained from the newspaper.
   c. are collected specifically for the current study.
   d. are less reliable than data from a secondary source.
   e. indicate correlation.

2. Data from secondary source
   a. are collected for other purposes than the current study.
   b. can be obtained from the newspaper.
   c. are collected specifically for the current study.
   d. are more reliable than data from a primary source.
   e. both a and b.

3. A census
   a. consists of information from all members of the population.
   b. consists of information from a subset of the population.
   c. is a secondary source of data.
   d. is less reliable than a secondary source of data.
   e. is less reliable than a primary source of data.

4. A sample
   a. consists of information from all members of the population.
   b. consists of information from a subset of the population.
   c. is a secondary source of data.
   d. is less reliable than a secondary source of data.
   e. is less reliable than a primary source of data.

5. If you are interested in how a football coach divides the practice session into drills, weight training, scrimmaging and game plans, it would be best to use a
   a. bar graph.
   b. line graph.
   c. pie chart.
   d. times series graph.
   e. component-part line chart.

6. If you are interested in how the earnings of a company have fluctuated over time, it would be best to use a
   a. bar graph.
   b. time series graph.
   c. pie chart.
   d. component-part line chart.
   e. histogram.

7. If you were interested in comparing the average earnings and interest expense for IBM with the average earnings and interest expense for Apple Computers, it would be best to use a
   a. bar graph.
   b. line graph.
   c. pie chart.
   d. time series graph.
   e. histogram.

8. In measuring the height of students, a systematic error could occur if
   a. the ruler used to measure students' height is one inch too long.
   b. we measure the height of the wrong students.
   c. we ask all students to remove their shoes.
   d. both male and female students are measured.
   e. we forget to record one student's height.

9. A random error could occur if
   a. the ruler used to measure students' height is one inch too long.
   b. we measure the height of the wrong students.
   c. we forget to have students remove their shoes.
   d. both male and female students are measured.
   e. we forget to record one student's height.

10. It is best to use a census when conducting a survey if
   a. the population is large.
   b. the population is small.
   c. we have a limited amount of time to conduct the survey.
   d. we would like to keep the costs of the survey low.
   e. the population is spread over a large geographic region.

11. It is best to use a sample when conducting a survey if
    a. the population is large.
    b. the population is small.
    c. we have a limited amount of time to conduct the survey.
    d. we would like to keep the costs of the survey low.
    e. all of the above except b.

12. Graphs can be useful for
    a. summarizing large amounts of data.
    b. showing trends in data.
    c. adding visual appeal to business reports.
    d. making comparisons.
    e. all of the above.

True/False (If false, explain why)

1. A golfer interested in knowing the percentage of total shots he hits with each different club should use a line graph.
2. Primary data are data collected specifically for the study.
3. Stock price data collected from the *Wall Street Journal* are primary data.
4. When the Nielsen television ratings collect data on television viewers to estimate the number of television viewers, they are using secondary data.
5. Random error is the difference between the value taken from a random sample and the value obtained from taking a census.
6. Systematic error results from problems in measurement.
7. Line charts are good for showing how the whole is divided into several parts.
8. Time series graphs show how data fluctuate over time.
9. The Gallup election poll uses primary data.
10. A component-part line chart can be useful for showing how gross national product is divided between consumption, investment, government spending, and net exports.
11. Time series graphs can be useful for examining trends in a company's financial ratios.
12. Financial ratios are often used in accounting and finance because they allow us to compare different size companies.

Questions and Problems

1. Suppose this month you spend $250 on rent, $125 on food, and $75 on entertainment. Use a pie chart to show how your money was spent.

2. Suppose the average earnings per share for ABC Company is $3 and the average earnings per share for XYZ Corporation is $6. Use a bar chart to compare the EPS for the two companies.

3. Below is the advertising budget for the Shady Lamp Shade Company from 1989 to 1993.

Year	T.V.	Radio	Newspaper	Total
1995	100	25	18	143
1996	105	31	27	163
1997	115	40	33	188
1998	123	42	34	199
1999	151	47	39	237

   Use a component-part line chart to show how the advertising budget has been divided between television, radio, and newspaper advertising over this five-year period.

4. Briefly explain why financial ratios are preferred to absolute numbers in financial analysis.

5. Below is a time series graph of the debt-equity ratios for the Cautious Electronics Company and the High Flying Electric Company. What conclusions can you draw by examining this graph?

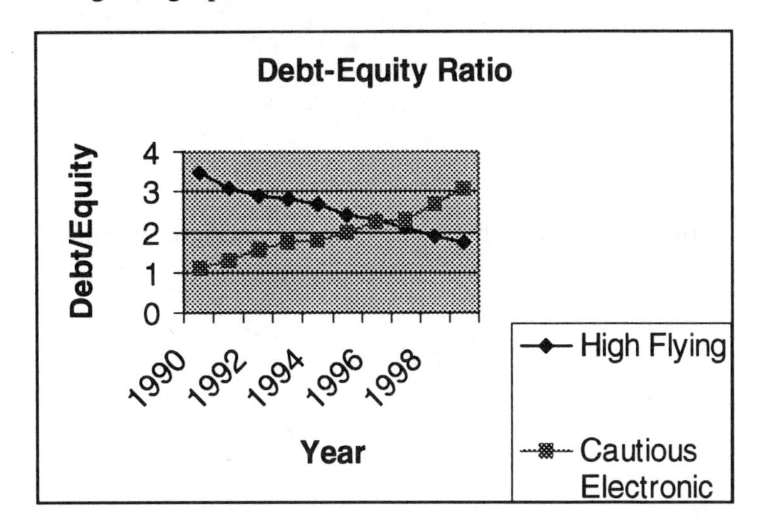

6. Below is a pie graph of the Smyth family's household budget. As their financial planner, what advice might you offer?

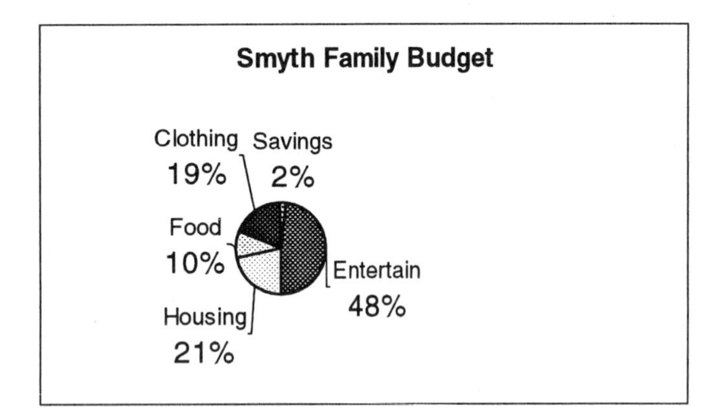

Answers to Supplementary Exercises

Multiple Choice

1.	c	6.	b	11.	e
2.	e	7.	a	12.	e
3.	a	8.	a		
4.	b	9.	b		
5.	c	10.	b		

True/False

1. False. A pie graph is best for showing how the "whole" is divided into parts.
2. True
3. False. Secondary data.
4. False. Primary data.
5. True
6. True
7. False. Pie chart.
8. True
9. True
10. True
11. True
12. True

Questions and Problems

1.

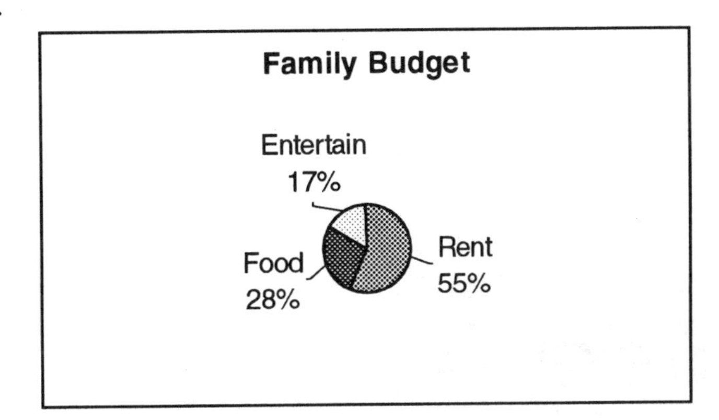

2.

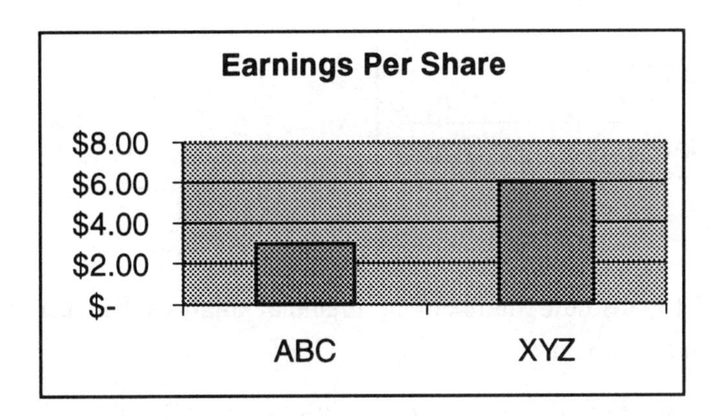

3.

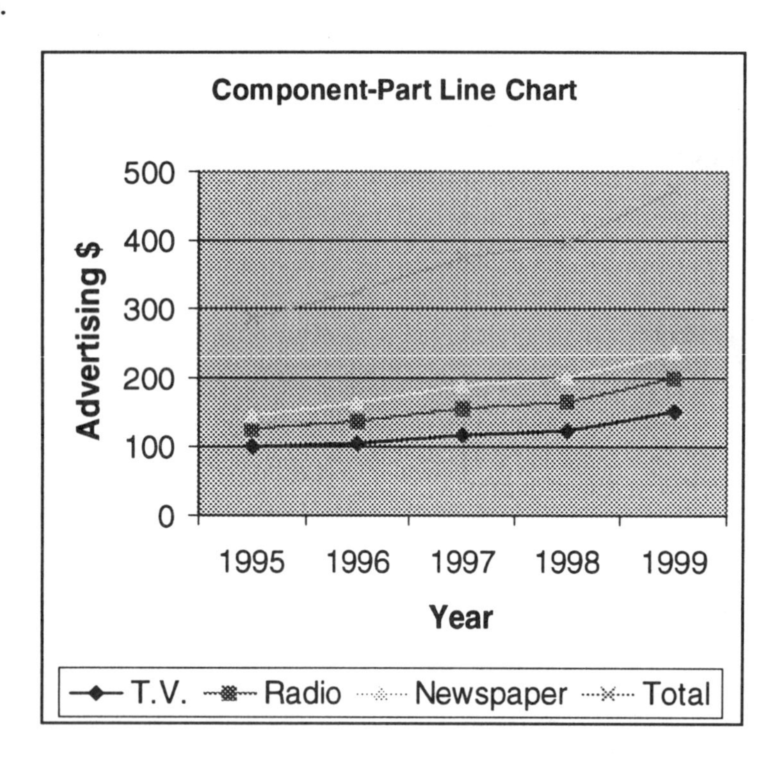

4. Ratios are generally preferred to absolute numbers in financial analysis because they allow us to compare different size companies.

5. From the graph, we can see that the two companies' debt-equity ratios have been moving in different directions over the last few years. High Flying has been reducing the amount of borrowing (debt) over recent years, whereas Cautious has been increasing its debt. Although the average debt-equity ratios for the two companies may be similar, it is clear that in terms of borrowing they are moving in opposite directions. Time series graphs are important because they allow us to compare trends in the financial ratios of different companies.

6. Looking at the Smyth family's budget, we can see that they spend a very high proportion of their budget on clothing and entertainment and put a very small amount into savings. As their financial planner, you might advise them to reduce their entertainment and clothing expenses and try to increase their savings.

CHAPTER 3

FREQUENCY DISTRIBUTIONS AND DATA ANALYSES

Chapter Intuition

Suppose you are taking a freshman economics class with 400 other students. Because of the large amount of data, a distribution of midterm grades would be very difficult to interpret. However, by grouping the grades into groups such as 90-100, 80-89, etc., it becomes much easier to evaluate the performance of the class. In addition, sometimes a simple table may not be very meaningful for presenting the data. In this case, we may want to organize the data by setting up classifications and then seeing how many observations or events fall into each classification. These types of tables are referred to as ***frequency tables***. Frequency tables can be useful for a teacher who wants to know how many students scored 90 percent or better on an exam or for a car dealer who wants to keep a tally of the number of cars sold each month.

Chapter Review

1. Once data have been collected, they need to be organized for easy analysis. In Chapter 2, we learned that data are much easier to analyze when presented in tables or graphs. However, when we have a large amount of data, simply placing each observation in a table may not be very useful. For example, if a teacher has 400 students, simply listing the midterm scores of all 400 students in a table may not help the teacher determine the class's performance. It would be more useful for the teacher to ***group the data***. The teacher might keep a count of the number of students who scored from 96-100, from 91-95, from 86-90, etc. This would make it much easier to see how the class performed.

2. Once the data are grouped, we might be interested in the number of observations that fall into each group. In our scores example, the teacher could use a ***tally table*** to keep track of the number of scores between 96 and 100, etc. The number of scores in each group is known as the ***frequency***. Any table that shows frequencies is called a ***frequency table***.

3. Sometimes we are not only interested in the number of observations in each group, but also in the number of observations in groups one and two combined. When we look at

the number of observations in both the first and second groups, we are looking at ***cumulative frequency***. Cumulative frequency is the number of occurrences in the group we are looking at and all the groups which came before. From our previous example, the teacher might be interested in the number of students who scored from 96 to 100 and from 91 to 95. In this case, he would be looking at the cumulative frequency for the first two groups.

4. Sometimes it is more interesting to look at the number of occurrences in each group as a proportion of all the data. When we do this, we are looking at ***relative frequency***. Again, from our example, the teacher might be interested in the number of students who received a score between 96 and 100 as a proportion of the entire class.

5. Sometimes we are interested in the proportion of occurrences in several groups. In this case, we would look at the ***cumulative relative frequency***. Like cumulative frequency, cumulative relative frequency keeps track of the total proportion of occurrences in the current group and all previous groups.

6. As we saw in Chapter 2, graphs can be a very effective way to present data. ***Histograms***, which are similar to the bar graphs presented in Chapter 2, can be used to present the frequency, cumulative frequency, relative frequency and relative cumulative frequency. The difference between histograms and bar graphs is that the neighboring bars touch each other and the area inside any bar is proportional to the number of observations in the corresponding class for a histogram.

7. A ***stem-and-leaf display*** is less of a graph and more of a table. In a stem-and-leaf display, the first digits of the data are presented in one column (the tree trunk) in ascending order. The remaining digit for each observation is presented to the right of the tree trunk and represents the leaves of the tree.

8. The ***Lorenz curve*** is a cumulative frequency curve which shows the distribution of a society's income. The ***Gini coefficient*** is used in conjunction with the Lorenz curve to express income distribution. If the Gini coefficient is equal to 0, we have perfect equality of income; that is, 20% of the population has 20% of the society's income, etc. If the Gini coefficient equals 1, then we have absolute inequality of income, that is, one family or person receives all of the society's income.

Example Problems

Example 1 **Constructing Tally Tables**

Suppose the midterm exam scores from Ms. Hannah's economics course are: 98, 95, 80, 72, 65, 90, 71, 55, 83, 88, 77, 79, 66, 45, 62.

Construct a tally table of midterm exams using the following intervals: 90-100, 80-89, 70-79, 60-69, 50-59, and less than 50.

Solution:

90-100 ///
80-89 ///
70-79 ////
60-69 ///
50-59 /
< 50 /

Example 2 **Frequency and Cumulative Frequency**

Use the tally table you constructed in example problem 1 to construct a frequency and cumulative frequency table.

Solution:

Frequency Table

Score	Frequency	Cumulative Frequency
90-100	3	3
80-89	3	6
70-79	4	10
60-69	3	13
50-59	1	14
< 50 1		15

Example 3 **Relative and Cumulative Relative Frequency**

Use your results from example problem 2 to construct a relative frequency and cumulative relative frequency table.

Solution:

Relative Frequency Table

Score	Frequency	Relative Frequency	Cumulative Relative Frequency
90-100	3	0.200	0.200
80-89	3	0.200	0.400
70-79	4	0.267	0.667
60-69	3	0.200	0.867
50-59	1	0.067	0.933
< 50 1		0.067	1.000
Total	15		

Example 4 **Frequency and Cumulative Frequency Histograms**

Use your results from example problem 2 to construct a frequency and cumulative frequency histogram.

Solution:

Frequency Histogram

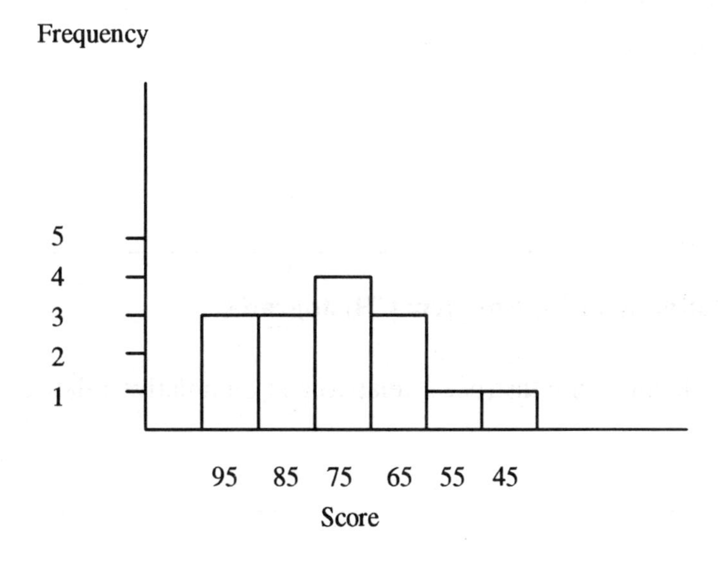

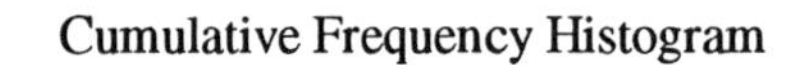

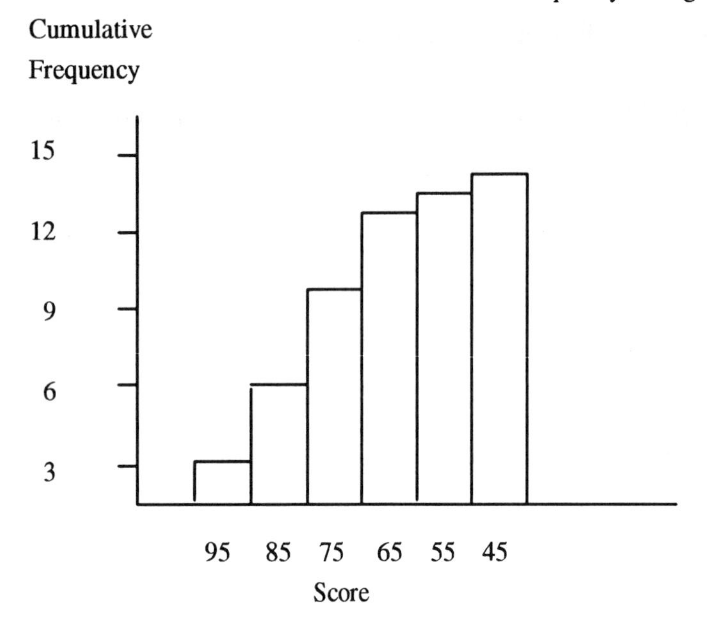

Example 5 **Relative and Cumulative Relative Frequency Histograms**

Use your results from exercise problem 3 to construct a relative and cumulative relative frequency histogram.

Solution:

Relative Frequency Histogram

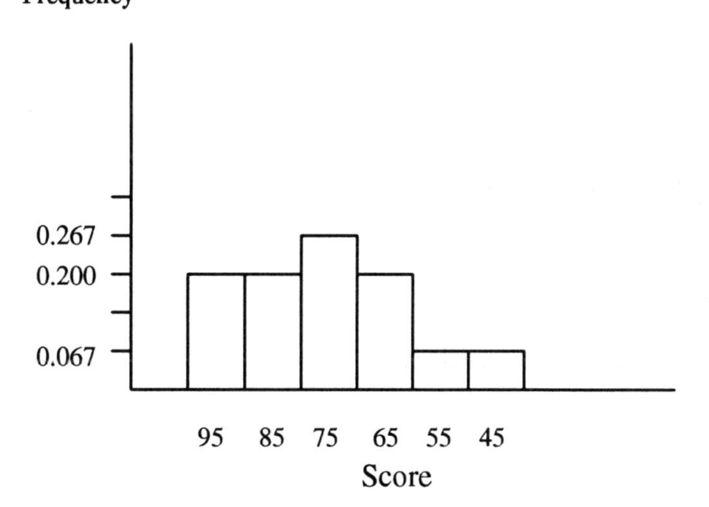

Cumulative Relative Frequency Histogram

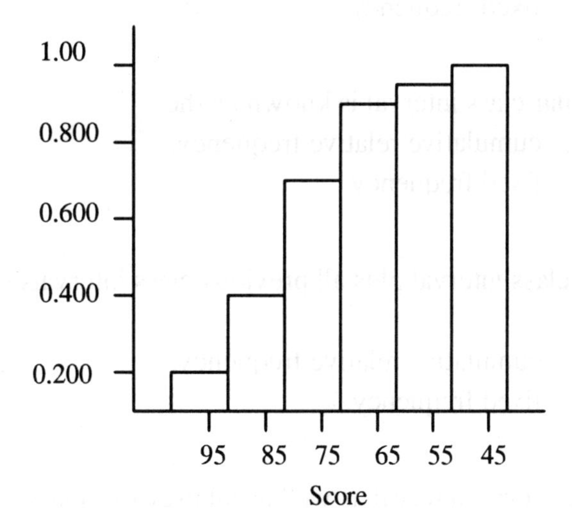

Supplementary Exercises

Multiple Choice

1. A tally table is
   a. used to organize raw data.
   b. the next step after constructing a frequency table.
   c. another name for a frequency histogram.
   d. similar to a stem-and-leaf display.
   e. an example of the Gini coefficient.

2. Data reported in a frequency table with class intervals is
   a. grouped data.
   b. ungrouped data.
   c. raw data.
   d. suitable primarily for academic use.
   e. always reported in a stem-and-leaf display.

3. The number of observations in a particular class interval is known as the
   a. frequency.
   b. relative frequency.
   c. cumulative frequency.
   d. cumulative relative frequency.
   e. fixed frequency.

4. The proportion of observations in a particular class interval is known as the
   a. frequency.
   b. relative frequency.
   c. cumulative frequency.
   d. cumulative relative frequency.
   e. fixed frequency.

5. The number of observations in the current class interval plus all previous class intervals is known as the
   a. frequency.
   b. relative frequency.
   c. cumulative frequency.
   d. cumulative relative frequency.
   e. fixed frequency.

6. The proportion of observations in the current class interval as well as all previous class intervals is known as the
   a. frequency.
   b. relative frequency.
   c. cumulative frequency.
   d. cumulative relative frequency.
   e. fixed frequency.

7. A stem-and-leaf display can be used for summarizing
   a. very large amounts of data.
   b. small amounts of data.
   c. any amount of data.
   d. grouped data.
   e. tally table results.

8. A frequency polygon is obtained by
   a. constructing a frequency histogram.
   b. constructing a cumulative frequency histogram.
   c. linking the midpoints from a frequency histogram.
   d. using a line graph.
   e. constructing a stem-and-leaf display.

9. The Lorenz curve is
   a. a specific type of histogram.
   b. a specific pie chart.
   c. a curve showing a society's distribution of income.
   d. a stem-and-leaf display.
   e. a frequency polygon.

10. When the Gini coefficient is equal to 1 there is
    a. absolute equality of income.
    b. absolute inequality of income.
    c. a 100% income tax.
    d. no income in the society.
    e. economic growth.

11. When the Gini coefficient is equal to 0 there is
    a. absolute equality of income.
    b. absolute inequality of income.
    c. a 100% income tax.
    d. no income in the society.
    e. economic growth.

12. Histograms are similar to bar graphs except
    a. neighboring bars do not touch each other.
    b. the area inside any bar is proportional to the number of observations in the corresponding class.
    c. the midpoints of the bars are connected.
    d. they are relative to the Gini coefficient.
    e. they are derived from frequency polygons.

True/False (If false, explain why)

1. Raw data are data that have been grouped into classes.
2. When the Lorenz curve is a straight line, there is be perfect equality of income.

3. It is usually best to limit the amount of data to 100 observations when using a stem-and-leaf display.
4. Grouping data makes handling a large data set less manageable.
5. Information may be lost when raw data are grouped.
6. Frequencies, cumulative frequencies, relative frequencies, and cumulative relative frequencies can all be graphed using histograms.
7. The area inside each bar of an histogram is proportional to the number of observations in that class.
8. A tally table is usually the first step in creating a frequency table.
9. When grouping data, the number of classes is unimportant.
10. Relative frequencies can be computed by adding together the frequencies of the current and all previous classes.
11. Cumulative frequencies measure the proportion of observations in a particular class.
12. A frequency polygon is obtained by linking the midpoints of the class intervals in a frequency histogram.
13. Cumulative frequencies can be useful for a teacher who would like to know the number of students receiving a grade of C or better.

Questions and Problems

1. Suppose the grades on a midterm exam in economics are: 95, 88, 92, 71, 64, 32, 89, 85, 90, 99, 72, 73, and 61. Construct a stem-and-leaf display for the data.
2. Use the data in Problem 1 to construct a tally table, with classes 91-100, 81-90, 71-80, 61-70, and 60 and below.
3. Use the tally table from Problem 2 to construct a frequency and cumulative frequency table.
4. Use the tally table in Problem 2 to construct a relative frequency and cumulative relative frequency table.
5. Draw a frequency and cumulative frequency histogram using the results from Problem 3.
6. Draw a relative and cumulative relative frequency histogram using your results from Problem 4.

Answers to Supplementary Exercises

Multiple Choice

1. a	6. d	11. b
2. a	7. b	12. b
3. a	8. c	
4. b	9. c	
5. c	10. a	

True/False

1. False. Raw data have not been grouped.
2. True
3. True
4. False. Grouping data makes a large data set more manageable.
5. True
6. True
7. True
8. True
9. False. Using too many classes defeats the purpose, whereas using too few classes limits the amount of information.
10. False. Relative frequencies can be computed by dividing the frequency in one class by the total number of observations in all classes.
11. False. Cumulative frequencies are the sum of observations in a particular class plus all preceding classes.
12. True
13. True

Questions and Problems

1.

Stem-and-Leaf Display

Stem	Leaf
9	0 2 5 9
8	5 8 9
7	1 2 3
6	1 4
5	
4	
3	2

2. Tally Table

Score	
91-100	///
81-90	////
71-80	///
61-70	//
< 60	/

3. Frequency and Cumulative Frequency Table

Score	Frequency	Cumulative Frequency
91-100	3	3
81-90	4	7
71-80	3	10
61-70	2	12
< 60	1	13

4. Relative and Cumulative Relative Frequency

Score	Relative Frequency	Cumulative Relative Frequency
91-100	0.231	0.231
81-90	0.308	0.538
71-80	0.231	0.769
61-70	0.154	0.923
< 60	0.077	1.000

5.

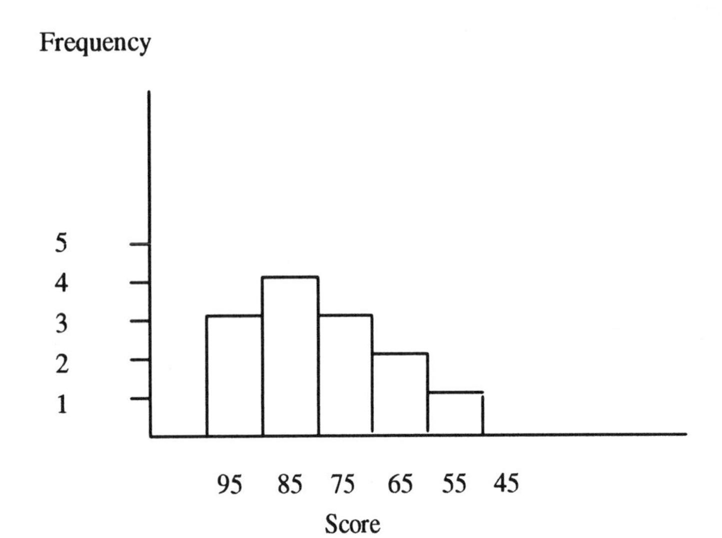
Frequency Histogram
Frequency
5
4
3
2
1
95 85 75 65 55 45
Score

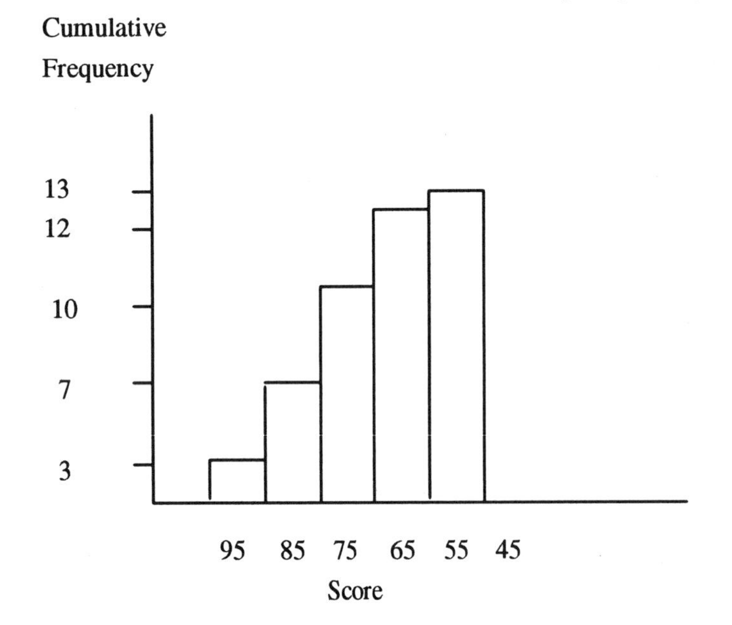
Cumulative Frequency Histogram
Cumulative
Frequency
13
12
10
7
3
95 85 75 65 55 45
Score

6.

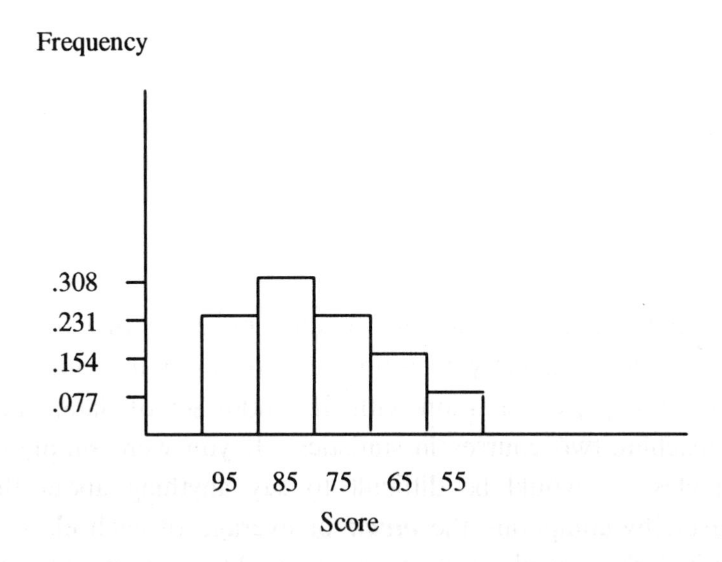
Relative Frequency Histogram
Frequency
.308
.231
.154
.077
95 85 75 65 55
Score

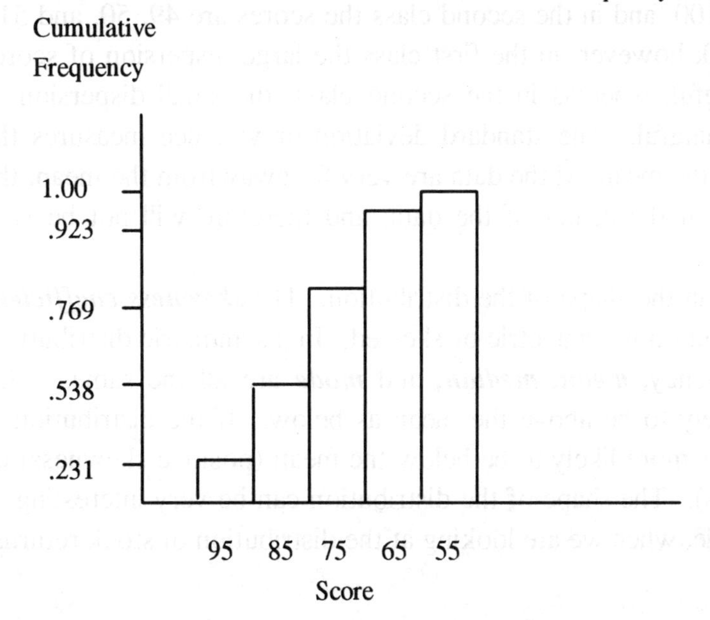
Relative Cumulative Frequency Histogram
Cumulative
Frequency
1.00
.923
.769
.538
.231
95 85 75 65 55
Score

CHAPTER 4

NUMERICAL SUMMARY MEASURES

Chapter Intuition

This chapter deals with ways to summarize a large amount of data by using one or two numbers that describe all the data. These summary measures can be used to compare different sets of data or to compare a single observation with the entire set of data. For example, suppose an instructor is teaching two courses in statistics. If you were simply to look at all the scores from each class, it would be difficult to say anything about the performance of each class. However, by computing the ***mean*** or average of each class, a simple comparison can be made. Similarly, a single student's grade could be compared with the class average to see how he or she is doing.

Once we have computed the average or center of the distribution, we are interested in knowing whether this number is meaningful. Measures of dispersion such as the ***variance*** and ***standard deviation*** allow us to determine if our measure of central tendency is meaningful. For example, suppose we have two classes each consisting of three students. In the first class, the midterm scores are 0, 50, and 100, and in the second class the scores are 49, 50, and 51. In both classes the mean will be 50; however, in the first class the large dispersion of scores makes the mean not very meaningful, whereas in the second class, the small dispersion of scores makes the mean very meaningful. The standard deviation or variance measures the deviation of each observation from the mean. If the data are very far away from the mean, the mean will not be a good measure of the center of the data, and therefore will not be very meaningful.

Finally, we are often interested in the shape of the distribution. The ***skewness coefficient*** allows us to determine if the distribution is symmetric or skewed. In a symmetric distribution, the three measures of central tendency, ***mean***, ***median***, and ***mode*** are all the same. This means that observations are as likely to be above the mean as below. If the distribution is skewed, an individual observation is more likely to be below the mean (positive skewness) or above the mean (negative skewness). The shape of the distribution can be very interesting in business and economics, for example, when we are looking at the distribution of stock returns.

Chapter Review

A series of data can be described using several descriptive statistics.

1. The ***mean***, ***median***, and ***mode*** are measures of the central tendency of the data. In its simplest terms, measures of central tendency give us an idea of where the average of the data is. The mean is the simple average of the data: the sum of the data divided by the number of observations. The median is the number in the middle. The mode is the most frequently repeating number. The median would be preferred to the other two when the data represent extreme values. For example, the mean salary of baseball players would be greatly influenced by the very large salaries of superstars such as Mark McGuire and Greg Maddox. The median would be a more meaningful figure. We would use the mode when we are interested in looking at the popularity of something. For example, a clothing store might be interested in the shoe size most often purchased. In this case, the mode would provide the store with information about the size of shoes that are in the greatest demand.

2. The ***variance***, ***standard deviation***, ***coefficient of variation***, and ***mean absolute deviation*** are measures of the dispersion of the data or how the data are spread out around their average. The variance measures the average squared distance of the observations from the mean. The standard deviation is just the square root of the variance. The mean absolute deviation measures the average distance the observations are from the mean. If we are interested in using descriptive statistics to compare data with different units, like the risk for two different stocks, we may use the coefficient of variation, which is unit free.

3. The shape of a distribution is described by its ***skewness*** or ***coefficient of skewness*** and by its ***kurtosis***. Skewness measures whether or not the data are symmetrical. Kurtosis measures the peakedness of the distribution.

4. When we have a large number of data, it is difficult to compute the mean and variance using every observation. For example, imagine computing the variance for 2,000 separate observations. In this case, it may be better to group the data, and deal with only 5 groups rather than 2,000 observations. The grouping procedure does not change the interpretation of the statistics, it only simplifies the computation process.

Useful Formulas

Measures of central tendency:

Sample arithmetic mean:

$$\bar{x} = \frac{\sum_{i=1}^{N} x_i}{N}$$

Sample geometric mean:
$\bar{x}_g = (x_1 \times x_2 \times x_3 \times \ldots \times x_N)^{1/N}$

Grouped mean:

$$\bar{x} = \frac{\sum_{i=1}^{k} f_i m_i}{\sum_{i=1}^{k} f_i}$$

Median:

$$m = L + \frac{(N/2 - F)}{f}(U - L)$$

Measures of dispersion:

Population variance:

$$\sigma^2 = \frac{\sum_{i=1}^{N} (x_i - \mu)^2}{N}$$

Sample variance:

$$s^2 = \frac{\sum_{i=1}^{N} (x_i - \bar{x})^2}{N - 1}$$

Mean absolute deviation:

$$MAD = \frac{\sum |x_i - \mu|}{N}$$

Coefficient of variation:

$CV = s / \bar{x}$

Population standard deviation:

$$\sigma = \sqrt{\frac{\sum_{i=1}^{N} (x_i - \mu)^2}{N}}$$

Sample standard deviation:

$$s = \sqrt{\frac{\sum_{i=1}^{N} (x_i - \bar{x})^2}{N - 1}}$$

Population variance for frequency distribution:

$$\sigma^2 = \frac{\sum_{i=1}^{k} f_i (m_i - \mu)^2}{N}$$

Sample variance for frequency distribution:

$$s^2 = \frac{\sum_{i=1}^{k} f_i (m_i - \bar{x})^2}{N - 1}$$

Measures of skewness:

Skewness:

$$\mu_3 = \frac{\sum_{i=1}^{N} (x_i - \mu)^3}{N}$$

Coefficient of skewness:

$$CS = \frac{\mu_3}{\sigma^3}$$

Example Problems

Example 1 **Computing the Average for Ungrouped Data**

Suppose you are given the following sales of 10 car dealers for last month. Compute the mean, median and mode.

48, 45, 67, 63, 38, 50, 58, 62, 40, 69

Solution:

mean = (48 + 45 + 67 + 63 + 38 + 50 + 58 + 62 + 40 + 69)/10
= 54

The median is the number in the middle. When the number of observations is an even number, the median is the mean of the middle two numbers. To find the median, we first rank the data from the smallest to the largest. The middle two numbers in our example are 50 (5th observation) and 58 (6th observation). Their mean is (50 + 58)/2 = 54.

The mode is the number that occurs most frequently. Since each number occurs only once in this problem, there is no mode. Note that the mean and median are equal to each other in this example merely by coincidence.

Example 2 **Computing Measures of Dispersion**

Using the information from example 1, let's compute the population variance, standard deviation, mean absolute deviation, and coefficient of variation.

Solution:

$$\begin{aligned}\sigma^2 &= [(48-54)^2 + (45-54)^2 + (67-54)^2 + (63-54)^2 \\ &\quad + (38-54)^2 + (50-54)^2 + (58-54)^2 + (62-54)^2 \\ &\quad + (40-54)^2 + (69-54)^2]/10 \\ &= 114\end{aligned}$$

$$\begin{aligned}\sigma &= \sqrt{\sigma^2} \\ &= \sqrt{114} \\ &= 10.67\end{aligned}$$

$$\begin{aligned}\text{MAD} &= [|48-54| + |45-54| + |67-54| + |63-54| \\ &\quad + |38-54| + |50-54| + |58-54| + |62-54| \\ &\quad + |40-54| + |69-54|]/10 \\ &= 9.8\end{aligned}$$

$$\begin{aligned}\text{CV} &= 10.67/54 \\ &= .1976\end{aligned}$$

Example 3 **Computing the Mean for Grouped Data**

Suppose we use the data from Example 1 and group the data from 36-40, 41-45, 46-50, 51-55, 56-60, 61-65 and 66-70.

Solution: First, we need to group the data and find the midpoints and frequency for each group.

Classes	Midpoints m_i	Frequency f_i	$f_i m_i$
36-40	38	2	76
41-45	43	1	43
46-50	48	2	96
51-55	53	0	0
56-60	58	2	116
61-65	63	2	126
66-70	68	1	68
			525

$$\text{Mean} = [(2 \times 38) + (1 \times 43) + (2 \times 48) + (0 \times 53) + (2 \times 58) + (2 \times 63) + (1 \times 68)]/10 = 52.5$$

$$\text{Median} = 56 + (10/2 - 5)/2\,(60 - 56) = 56$$

Example 4 **Computing Measures of Dispersion for Grouped Data**

Compute the variance, standard deviation, and coefficient of variation for the grouped data of Example 3.

Solution:

$$\sigma^2 = [2(38 - 52.5)^2 + 1(43 - 52.5)^2 + 2(48 - 52.5)^2 + 0(53 - 52.5)^2 + 2(58 - 52.5)^2 + 2(63 - 52.5)^2 + 1(68 - 52.5)^2]/10 = 107.25$$

$$\sigma = \sqrt{\sigma^2} = 10.36$$

$$CV = 10.36/52.5$$
$$= .1973$$

Example 5 **Measuring the Shape of the Data - Skewness**

Use the data from Example 1 to compute the coefficient of skewness. How are the data skewed?

Solution: First we need to compute μ_3.

$$\mu_3 = [(48-54)^3 + (45-54)^3 + (67-54)^3 + (63-54)^3 + (38-54)^3 + (50-54)^3 + (58-54)^3 + (62-54)^3 + (40-54)^3 + (69-54)^3]/10$$
$$= -88.36$$

$$CS = \mu_3/\sigma^3$$
$$= -88.36/1214.77$$
$$= -.0727$$

This result, being a negative number, means the data are slightly skewed to the left.

Supplementary Exercises

Multiple Choice

1. If the data is positively skewed, which of the following statements are true?
   a. The mean is greater than the mode.
   b. The mean is less than the mode.
   c. The mean and mode are the same.
   d. The mean, median, and mode are the same.
   e. The mode will be 0.

2. When comparing the geometric mean and arithmetic mean
   a. the geometric mean and the arithmetic mean will always give the same numeric value.
   b. the arithmetic mean will always be less than or equal to the geometric mean.
   c. the arithmetic mean will always be greater than or equal to the geometric mean.
   d. the mode will always equal the geometric mean.
   e. the mode will always equal the arithmetic mean.

3. The larger the variance is for a set of data, the
   a. more meaningful the mean is.
   b. less meaningful the mean is.
   c. variance plays no part in determining whether the mean is meaningful.
   d. smaller the standard deviation.
   e. smaller the coefficient of variation.

4. Kurtosis measures the
   a. center of the data.
   b. dispersion of the data.
   c. peakedness of the data.
   d. symmetry of the data.
   e. mode of the data.

5. The mode represents
   a. the most frequently repeating score.
   b. middle score.
   c. a geometric average.
   d. an arithmetic average.
   e. a combination of the geometric and arithmetic averages.

6. Which of the following statements is true?
   a. The mean, median, and mode will always be the same for a data set.
   b. The mean, median, and mode will never be the same for a set of data.
   c. The mean will always be greater than the median, but smaller than the mode.
   d. The mean can never be negative.
   e. The variance can never be negative.

7. Which of the following statements is true?
   a. The variance will always be larger than the mean absolute deviation.
   b. The variance will always be smaller than the mean absolute deviation.
   c. The variance and mean absolute deviation will be the same.
   d. The standard deviation can never be negative.
   e. The standard deviation can be negative.

8. Grouping data can
   a. enrich the meaning of the mean and standard deviation.
   b. simplify the computational process for the mean and standard deviation.
   c. change the interpretation of the mean and standard deviation after they are computed.
   d. complicate the computational process for the mean and standard deviation.
   e. not be used for large amounts of data.

9. The range represents the
   a. difference between the highest and lowest value.
   b. middle number.
   c. most frequently repeating number.
   d. highest number.
   e. lowest number.

10. A *z*-score will always have
   a. a mean of 1 and a standard deviation of 0.
   b. a mean of 1 and a standard deviation of 1.
   c. a mean of 0 and a standard deviation of 1.
   d. a mean of 0 and a standard deviation of 0.
   e. a mean of 0 and a standard deviation that is greater than 1.

11. When the variance for a set of data equals 0,
   a. half the numbers will lie above the mean and half below.
   b. the mean will always be 0.
   c. all numbers in the data set will be the same.
   d. the data will be widely dispersed around the mean.
   e. the standard deviation will be 1.

12. Suppose a teacher records the following scores for a test: 87 42 55 37 99 98 47. The median is
   a. 37
   b. 98
   c. 55
   d. 87
   e. 99

13. Suppose the weights of the linemen of the San Francisco 49ers are: 272, 291, 285, 272, 280, and 245. The mode is
   a. 272
   b. 291
   c. 285
   d. 280
   e. 245

14. Suppose a teacher finds that the midterm scores in her accounting class have a mean of 82 and a variance of 9. The standard deviation of the midterms is
    a. 82
    b. 9
    c. 2
    d. 9
    e. 9^2

True/False (If false, explain why)

1. Grouping data changes the interpretation of the mean.
2. The standard deviation is a more popular measure of dispersion than the variance because it is in the same units as the mean.
3. The coefficient of variation should not be used to compare the dispersion of different sets of data that are measured in different units.
4. Kurtosis measures the symmetry of the distribution.
5. Positive skewness of a stock's returns is desirable.
6. If the mean, median, and mode all equal 10, the data is positively skewed.
7. The larger the variance, the more meaningful the mean.
8. Skewness measures the peakedness of the distribution.
9. The variance can never be negative.
10. The standard deviation can be negative.
11. The coefficient of variation can never be negative.
12. The coefficient of variation is always positive.

Questions and Problems

1. You are given the following grades from a midterm exam in English: 95, 34, 83, 92, 94, 88, 99.
   Find the median and range of the scores.

2. You are given the following two groups of numbers

   1,000 2,000 3,000
   and 1,000,000 2,000,000 3,000,000
   Is it true that the second group of numbers has a higher dispersion because it has a higher standard deviation?

3. Compute the population mean and variance for the following numbers: 92, 94, 86, 42, 38, 99.

4. Use the data from Exercise 3 to compute the skewness coefficient. Are the data skewed?

5. When examining the desirability of a business venture, we sometimes use the variance of the profits to measure the risk of the project. Briefly explain why the variance may not be a good measure of business risk.

6. Find the median, mode and standard deviation for the following observations:
   32, 54, 88, 27, 99, 13, 72, 88, 90, 100, 21, 225

7. Compute the skewness coefficient for the data in Exercise 6.

8. Fill in the missing values in the table and find the mean for the frequency distribution.

Class	Midpoint m_i	Frequency f_i	$f_i m_i$
1-5	2.5	6	
6-10	7.5	3	
11-15	12.5	7	
16-20	17.5	2	
21-25	22.5	12	

9. Below are the dollar amounts spent on beer each month by 28 students at McBud University: $2, 7, 25, 18, 19, 21, 38, 30, 27, 19, 22, 31, 35, 29, 33, 22, 17, 18, 39, 33, 21, 3, 6, 24, 33, 28, 9, 17.

   a. Group the data, using groups of 0-9, 10-19, 20-29, and 30-40.

   b. Use a stem-leaf graph to show the data.

   c. Compute the mean and variance using the grouped data.

Answers to Supplementary Exercises

Multiple Choice

1. b	6. e	11. c
2. c	7. d	12. c
3. b	8. b	13. a
4. c	9 a	14. d
5. a	10. c	

True/False

1. False. Grouping data does not change the interpretation of the mean, it only simplifies the computation.
2. True
3. False. The coefficient of variation should be used to compare different data sets because it is unit free.
4. False. Kurtosis measures the peakedness of the distribution.
5. True
6. False. The data are symmetric.
7. False. A large variance makes the mean less meaningful.
8. False. Skewness measures symmetry.
9. True
10. False. The standard deviation can never be negative.
11. False. If the mean is negative, the coefficient of variation will be negative.
12. False. See answer to 11.

Questions and Problems

1. Median = 92
   Range = 65

2. No. Because the two sets of data are in different units, it is not appropriate to compare their standard deviations. If we were to use the coefficient of variation as our measure of dispersion, the dispersion would be the same for both sets of data.

   $\bar{x} = 75.167$

3. $\sigma^2 = 634.139$

4. CS = –.6426

5. When we use variance as a measure of risk, we consider values above the mean to be as undesirable as values below the mean. Because profits that lie above the mean are desirable, rather than risky, the variance may not be an appropriate measure of business risk.

6. Median = 80
   Mode = 88
   Standard deviation = 54.48

7. 31.69

8.

Classes	Midpoints m_i	Frequency f_i	$f_i m_i$
1-5	2.5	6	15
6-10	7.5	3	22.5
11-15	12.5	7	87.5
16-20	17.5	2	35
21-25	22.5	12	270
		30	430

Mean = 430/30 = 14.3

9. a.

Class	Frequency f_i	Midpoint m_i	$f_i m_i$
0-9	5	4.5	22.5
10-19	6	14.5	87.0
20-29	9	24.5	220.5
30-39	8	34.5	276.0

b. Stem-leaf table

Stem	Leaves
0	2 3 6 7 9
10	7 7 8 8 9 9
20	1 1 2 2 4 5 7 8 9
30	0 1 3 3 3 5 8 9

c. Mean = 606/28 = 21.64

Variance = 113.26

CHAPTER 5

PROBABILITY CONCEPTS AND THEIR ANALYSIS

Chapter Intuition

With the exception of death and taxes, life is full of uncertainty. ***Probability*** is a counting method used to help us quantify this uncertainty. For example, suppose we have a bag filled with five balls, three white and two green. If you are asked what is the chance of picking a green ball out of the bag, your intuition would tell you that you have two chances out of five or a 40% chance. What you have done is compute a probability by dividing the number of favorable outcomes (green balls) by the total possible outcomes (five possible balls to select).

Although the previous example may be quite obvious, it is the basic building block for probability and statistics. Using this same counting method, we can consider a more difficult problem. Suppose you are now interested in knowing the probability of selecting a green ball on your second draw given that the first ball you selected was white and it was not placed back in the bag. Again, your intuition would tell you that there are now four balls in the bag and two are white, so the chance of selecting a white ball is 50%. What you have just done is compute a conditional probability, that is, the probability of some event occurring (choosing a green ball on the second draw) given that the first ball drawn was white.

Although these examples may be obvious, more complicated problems will require rules that help us to decide how to count the possible outcomes and how to determine the probability. The rules of counting will depend on whether the events are related or independent (unrelated). There are different ways to compute probability: 1) use intuition to determine the ratio of favorable outcomes to total possible outcomes, 2) use established rules, such as ***conditional probability***, rules of addition and multiplication, 3) use established probability distributions. This chapter relies on methods of counting and on the established rules of probability. Chapter 6 will show you how to use established probability distributions.

Chapter Review

1. Probabilities can take on values between 0 (no chance of the event occurring) to 1 (event occurs with certainty). In ***classical probability***, we look at how many favorable

outcomes can occur as a percentage of the total number of outcomes. For example, the probability of tossing a coin and receiving a head is 50%, since there is a total of only two possible outcomes, a head or a tail.

2. A ***union of two events*** occurs when either Event A or B occurs. An ***intersection of two events*** occurs when both Event A and Event B occur. For example, if we choose one card from a deck of cards, we might be interested in the chance of receiving a face card (Event A) and/or the chance of receiving a club (Event B). The union of Event A and Event B would deal with the chance of drawing a face card or a club, while the intersection of Event A and B would deal with the chance of drawing a face card that was also a club.

3. ***Joint probability*** is the probability that two or more events will occur together. For example, the probability that a Republican wins the presidential election and the Republicans gain control of the Senate is a joint probability.

4. ***Marginal probability*** represents a way of separating a joint probability into ***unconditional probabilities***. From our previous example, we might be interested only in the probability that a Republican wins the presidential election, regardless of who has control of the Senate.

5. ***Conditional probability*** is the probability of some event B occurring, given that some other event A has already occurred. For example, we might be interested in the probability of Horse A winning a race given a muddy track.

6. ***Bayes' Theorem*** allows us to revise our probabilities by using new information.

7. The concepts of ***permutations*** and ***combinations*** provide important tools for computing probabilities. A popular example of probability based on combinations is the birthday problem, where we are interested in the probability that at least two people in a given room have the same birthday. As we increase the number of people in the room the number of possible combinations increases. With only two people in a room, there is only one chance for a match. With three people in a room, there are now three possible matches: Person A with B; A with C; and B with C. With four people in a room there are six possible matches and so on. With only 50 people in a room, there is a 97% chance that at least two people have the same birthday. This result comes from the large number of combinations of people who could have the same birthday in a room of 50 people.

Useful Formulas

Probability of Event A:

$$P_r(A) = \frac{\text{Number favorable outcomes}}{\text{Total number of outcomes}}$$

Union of Events A and B:

$$P_r(A \cup B) = P_r(A) + P_r(B) - P_r(A \cap B)$$

Intersection of Events A and B:

$$P_r(A \cap B) = P_r(A) + P_r(B) - P_r(A \cup B)$$

Probability of Complements:

$$P_r(A \cup \overline{A}) = P_r(A) + P_r(\overline{A}) = 1$$

Conditional Probability:

$$P_r = (A|B) = \frac{P_r(A \cap B)}{P_r(B)}$$

Bayes' Theorem:

$$P_r = (A|B) = \frac{P_r(B|A)P_r(A)}{P_r(B)}$$

Number of Permutations of n-things taken r at a time

$${}_nP_r = \frac{n!}{(n-r)!}$$

Number of Combinations r objects can be selected from n

$${}_nP_r = \frac{n!}{r!(n-r)!}$$

Example Problems

Example 1 **Independent and Mutually Exclusive Events**

In each situation below, decide whether the two events are independent, mutually exclusive, or neither.

a. The Detroit Lions win the Super Bowl in 1999 and the Seattle Seahawks win the Super Bowl in 1999.

b. The Boston Celtics win the NBA Championship and the Celtics sell more than 1 million tickets in the same season.

c. A person wins both the New York and New Jersey lotteries.

d. You get a head in one flip of a coin and get a tail in another flip of a coin.

Solution: Events are mutually exclusive if, when one event occurs, the other event cannot occur. Events are independent if the chance of one event occurring is not affected by the other event.

a. Since both the Lions and the Seahawks cannot win the Super Bowl in 1999, these events are mutually exclusive.

b. These two events are not mutually exclusive since both could occur. These events are also not independent, since a winning Celtics team will probably influence ticket sales positively.

c. The events are not mutually exclusive since a person could win both lotteries. The events are independent because the probability of winning one lottery does not affect the probability of winning the other.

d. The events are not mutually exclusive. The events are independent since the probability of flipping a tail is not affected by receiving a head in the first toss.

Example 2 **Computing Probability With Replacement**

A bag contains five balls, 3 red and 2 white. If you draw 3 balls out of the bag one at a time with replacement, what is the probability that you receive a red ball first, a white ball second and a red ball third?

Solution: The probability of drawing a red ball first is 3 out of 5 or .60; the probability of drawing a white ball second, which is independent of the first draw because the ball has been replaced, is 3 out of 5 or .60; the probability of drawing a red ball on the third is 2 out of 5 or .40. So the probability that this occurs is,

$$.60 \times .40 \times .60 = .144 \text{ or } 14.4\%$$

Example 3 **Computing Probability Without Replacement**

Suppose we consider Example 2 again, except this time the balls are removed and not placed back in the bag. What is the probability that you receive a red ball first, a white ball second and a red ball third?

Solution: When the balls are not replaced in the bag, we can see that the probability of drawing a white ball second is conditional on drawing a red ball first. The probability of drawing a red ball first is 3 out of 5 or .60; the probability of drawing a white ball out second, given that a red ball has already been drawn is 2 out of 4 or .50; the probability of drawing a red ball on the third draw, given that a red ball and a white ball have already been drawn, is 2 out of 3 or .67. So the probability that this entire sequence occurs is

$$P_r(R_1 \cap W_2 \cap R_3) = P_r(R_1)\, P_r(W_2 \mid R_1)\, P_r(R_3 \mid R_1 \cap W_2)$$
$$= .60 \times .50 \times .67 = .201 \text{ or } 20.1\%$$

Example 4 **Conditional Probability**

$P_r(E1) = 0.5$, $P_r(E2) = 0.4$. Obtain $P_r(E1 \cup E2)$, $P_r(E1 \mid E2)$, and $P_r(E2 \mid E1)$ given that $P_r(E1 \cap E2) = 0.2$.

Solution:

$$\begin{aligned} P_r(E1 \cup E2) &= P_r(E1) + P_r(E2) - P_r(E1 \cap E2) \\ &= .5 + .4 - .2 \\ &= .7 \end{aligned}$$

$$\begin{aligned} P_r(E1 \mid E2) &= P_r(E1 \cap E2)/P_r(E2) \\ &= .2/.4 \\ &= .50 \text{ or } 50\% \end{aligned}$$

$$\begin{aligned} P_r(E2 \mid E1) &= P_r(E1 \cap E2)/P_r(E1) \\ &= .2/.5 \\ &= .40 \text{ or } 40\% \end{aligned}$$

Example 5 **Permutations**

Suppose you are at the racetrack and would like to play the exacta, in which you try to select the first and second place horse in correct order. If there are 8 horses in the field, what is the probability of winning the bet?

Solution: Once we have selected the two horses to finish first and second, we can compute the number of ways the other six horses can finish, which is 6!. Since there are 8! ways that these 8 horses can finish, we know $6!/8! = 1/56$ is the answer.

Example 6 **Conditional Probability**

Consider Example 5 again, except compute the probability of winning using conditional probability.

Solution: The probability that horse A finishes first is 1 out of 8. The probability that horse B finishes second given that horse A finishes first is 1 out of 7. Therefore, the probability that we win our bet is

$$
\begin{aligned}
P_r(A_1 \cap B_2) &= P_r(A_1)\, P_r(B_2 \mid A_1) \\
&= 1/8 \times 1/7 = 1/56
\end{aligned}
$$

Example 7 **Classical Probability**

If you roll a die twice, what is the probability that the sum of the rolls equals 5?

Solution: Each roll has a 1/6 probability and the two rolls are independent, so the probability of rolling a certain combination is 1/36 (that is, 1/6 x 1/6). In the first case, the possible combinations that will sum to 5 are: 1,4; 2,3; 3,2; and 4,1. So the probability of two rolls summing to 5 is just the sum of the probabilities that will add to 5:

$$
\begin{aligned}
P_r(x = 5) &= 1/36 + 1/36 + 1/36 + 1/36 \\
&= 1/9
\end{aligned}
$$

Example 8 **Union and Intersection of Events**

Consider the roll of a six sided die. Given the following events *A* and *B*, give the union and the intersection for *A* and *B* in each case.

a. A = {2, 4, 6} B = {1, 3, 5}

b. A = {2, 4} B = {1, 2, 4}

c. A = {1, 2, 3} B = {1, 2, 3}

d. A = {2, 3, 4, 5} B = {1, 4, 5, 6}

Solution: The union of A and B represents all elements that are in A or B. The intersection of A and B represents all elements that are in both A and B.

a. $A \cup B = \{1, 2, 3, 4, 5, 6\}$
$A \cap B = \{\ \}$

b. $A \cup B = \{1, 3, 5\}$
$A \cap B = \{2, 4\}$

c. $A \cup B = \{1, 2, 3\}$
$A \cap B = \{1, 2, 3\}$

d. $A \cup B = \{1, 2, 3, 4, 5, 6\}$
$A \cap B = \{4, 5\}$

Example 9 **Combinations**

Mary Gop, political candidate for governor has decided to campaign in 4 of the 7 largest cities in the state. How many different combinations of 4 cities can she stop in?

Solution: $_nC_r = n!/[r!(n - r)!]$
$= 7!/[4!(7 - 4)!]$
$= 35$

Example 10 **Marginal and Joint Probability**

Suppose 200 students were randomly asked whether they own a stereo and/or a personal computer. The results are given in the following table.

	PC (A)	No PC ($\bar{A}$)	
(B) Stereo	40	60	100
(B) No stereo	20	80	100
	60	140	200

Obtain $P_r(A \mid B)$, $P_r(A \cup B)$ and $P_r(B)$

Also, show that $P_r(A \mid B) = P_r(A \cap B)/P_r(B)$

Solution:

$$P_r(A|B) = \frac{P_r(A \cap B)}{P_r(B)} = \frac{40/20}{100/200} = 0.4$$

$$P_r(A \cap B) = 40/200 = 0.2$$

Example 11 **Bayes' Theorem**

A factory procures headlamps from two suppliers. Supplier A provides 40% of the headlamps and Supplier B provides 60%. Suppose 10% of the headlamps delivered by Supplier A and 20% of the headlamps delivered by Supplier B are bad. If a bad headlamp is found, what is the probability that it comes from Supplier B?

Solution: If we select a headlamp randomly from the factory, there is a 40% chance that it comes from Supplier A and a 60% chance that it comes from Supplier B. Therefore, the prior probabilities are

$P_r(A) = 40\% \quad P_r(B) = 60\%$

Once we have selected a bad headlamp, we now have additional information that we can use. We know that with a headlamp from Supplier A, there is a 10% chance that it is bad, $P_r(\text{bad} \bullet A) = 10\%$. Similarly, we know $P_r(\text{bad}|B) = 20\%$. Using this information we can solve for

$$P_r(B \mid \text{bad}) = \frac{P_r(B \cap \text{bad})}{P_r(\text{bad})} = \frac{P_r(\text{bad} \mid B)\ P_r(B)}{P_r(\text{bad} \mid B)\ P_r(B) + P_r(\text{bad} \mid A)\ P_r(A)}$$

$$= \frac{20\% \times 60\%}{20\% \times 60\% + 10\% \times 40\%} = \frac{12}{12 + 4} = \frac{3}{4}$$

Example 12 **Bayes' Theorem**

In Example 11, why is the posterior distribution $P_r(B \mid \text{bad})$ higher than $P_r(B)$?

Solution: Before we know the headlamp is bad, we know the probability that a headlamp comes from Supplier B is 60%. Because Supplier B has a higher failure rate than Supplier A, this increases still further the probability that a headlamp from Supplier B is bad.

Supplementary Exercises

Multiple Choice

1. The union of two events occurs when
   a. neither event A nor event B occur.
   b. both event A and event B occur.
   c. either event A or event B occur.
   d. event A and event B are mutually exclusive.
   e. event A and event B are independent.
2. The intersection of two events occurs when
   a. neither event A nor event B occur.
   b. both event A and event B occur.
   c. either event A or event B occur.
   d. event A and event B are mutually exclusive.
   e. event A and event B are independent.
3. If events A and B are independent events
   a. they occur simultaneously.
   b. event B is not conditional on event A occurring.
   c. they are not influenced by one another.
   d. they cannot occur simultaneously.
   e. they are mutually exclusive.
4. If events A and B are mutually exclusive
   a. they occur simultaneously.
   b. they cannot occur simultaneously.
   c. they are not influenced by one another.
   d. they are independent.
   e. event B is conditional on event A occurring.
5. Bayes' Theorem
   a. gives the probability of two events occurring jointly.
   b. gives the marginal probability of x.
   c. is a method for updating probabilities by using new information.
   d. is a method for constructing Venn diagrams.
   e. is a method for computing probabilities for mutually exclusive events.

6. The conditional probability of x given y is
   a. the probability that x and y occur jointly.
   b. the probability that y occurs if x has already occurred.
   c. the probability that x occurs if y has already occurred.
   d. the marginal probability of x minus the marginal probability of y.
   e. the same as a joint probability.

7. The probability of tossing three heads in a row is
   a. 1/2
   b. 1/4
   c. 1/8
   d. 1/6
   e. 1/16

8. The probability of receiving one head and one tail in two flips of a coin is
   a. the same as the probability of tossing two heads in a row.
   b. is less than the probability of tossing two heads in a row.
   c. is greater than the probability of tossing two heads in a row.
   d. is 0.
   e. is 1/2.

9. The multiplication rule in probability is only appropriate when the events
   a. are independent.
   b. are mutually exclusive.
   c. are dependent.
   d. occur jointly.
   e. are conditional upon one another.

10. If you roll two dice, the probability that the sum of the two rolls equals 7 is
    a. 1/36
    b. 2/36
    c. 1/6
    d. 3/36
    e. 1/4

11. If you roll two dice, the probability that the sum of the two rolls equals 12 is
    a. 1/36
    b. 2/36
    c. 1/6
    d. 3/36
    e. 1/4

12. Suppose the probability that you will be elected president of the student body is .30, and the probability that you will get into Harvard Law School is .05. If the two events are independent, the probability that you will not only become student body president but also get into Harvard Law School is
    a. .35
    b. .25
    c. .015
    d. .50
    e. .05

13. Suppose a bag is filled with 8 balls, 3 white and 5 black. If balls are drawn from the bag without replacement, the probability that the second ball drawn is black, given that the first ball drawn is black, is
    a. 1/8
    b. 5/8
    c. 4/7
    d. 2/8
    e. 1/5

14. Suppose a bag is filled with 8 balls, 3 white and 5 black. If balls are drawn from the bag with replacement, the probability that the second ball drawn is black is
    a. the same as the probability that the first ball drawn is black.
    b. greater than the probability that the second ball drawn is black.
    c. less than the probability that the second ball drawn is black.
    d. less than the probability that the second ball is white.
    e. less than the probability that the first ball is white.

15. Suppose you draw 1 card from a deck of 52 cards and flip a coin once. The probability that you draw a heart and toss a tail is
    a. $1/4 + 1/2$
    b. $1/4 \times 1/2$
    c. $1/52 \times 1/2$
    d. $1/52 + 1/2$
    e. $1/4 - 1/2$

True/False (If false, explain why)

1. $P_r(A)\, P_r(B) = P_r(A \cap B)$
2. $P_r(A) + P_r(B) = P_r(A \cap B)$
3. $P_r(A) + P_r(B) = P_r(A \cup B)$
4. $P_r(A \mid B) = P_r(A)$
5. $P_r(A \cap B) = P_r(B \cup A)$
6. $P_r(B \cup A) \geq P_r(A) + P_r(B)$
7. $P_r(B \cap A) \leq P_r(A) + P_r(B)$
8. Because there is a 50% probability of tossing a head and a 50% probability of tossing a tail, there is a greater probability of tossing a head on the first toss and a tail on the second toss than there is of tossing two heads in a row.
9. Bayes' Theorem is a method that allows us to update probabilities using new information.
10. The probability that two mutually exclusive events occur simultaneously is 1.

Questions and Problems

1. Suppose you toss a coin twice. What is the probability of tossing two heads?
2. Suppose you toss a coin twice. What is the probability of tossing two heads, given that your first toss was a head?
3. Suppose you are on vacation in Europe and would like to visit 11 cities. Unfortunately, you only have time to visit 5 of these cities. How many different combinations of 5 cities can you visit?
4. Suppose you draw four cards from a standard 52-card deck. What is the probability that the first card will be a club, the second a diamond, the third a heart and the fourth a spade if the cards are drawn with replacement?
5. How would your answer to Exercise 4 change if the cards were drawn without replacement?
6. Suppose an auto dealership is interested in the types of options people purchase on cars. The records of 125 people who purchased cars were examined to see which ones purchased air conditioning and which ones purchased automatic transmission. The results are given in the following table.

	A/C (A)	No A/C (A)	
(B) Automatic	63	10	73
(B) No Automatic	25	27	52
	88	37	125

   Find $P_r(B \mid A)$, $P_r(A)$, $P_r(B)$ and $P_r(A \cap B)$

7. A Senate committee consists of 12 Democrats and 9 Republicans. In how many ways can a subcommittee consisting of 8 members, 5 Democrats and 3 Republicans be formed?

Answers to Supplementary Exercises

Multiple Choice

1.	c	6.	c	11.	a
2.	b	7.	c	12.	c
3.	c	8.	c	13.	c
4.	b	9.	a	14.	a
5.	c	10.	c	15.	b

True/False

1. False. True if A and B are independent.
2. False. True if A and B are mutually exclusive.
3. False. True if A and B are mutually exclusive.
4. False. True if A is independent of B.
5. False. True if $P_r(A) = P_r(B) = 0$.
6. True
7. True
8. False. The probabilities are the same.
9. True
10. False. The probability that two mutually exclusive events occur simultaneously is 0.

Questions and Problems

1. 1/4
2. 1/2
3. 462
4. 1/256
5. $1/4 \times 13/51 \times 13/50 \times 13/49$

6. $P_r(B \mid A) = .716$
   $P_r(A) = 88/125$
   $P_r(A \cap B) = 63/125$
7. 41,580

CHAPTER 6

DISCRETE RANDOM VARIABLES AND PROBABILITY DISTRIBUTIONS

Chapter Intuition

In Chapter 5, you were introduced to the concept of probability. Probability theory presents a systematic approach for counting and assigning chance to uncertain outcomes. In this chapter, you will see that data generated in certain ways will have unique distributions. This can make it easier to compute probabilities or parameters such as the mean and variance. For example, if we flip a coin 100 times we would be able to create a series of data consisting of H (heads) or T (tails). The series of data we have just created will have a unique pattern or distribution that can be useful for calculating probability. This distribution is known as the binomial distribution. Statisticians have already established formulas for computing the probability associated with the outcomes from the binomial distribution.

Although the tossing of a coin may seem like a useless and trivial exercise, there are many types of real world examples that follow a similar pattern. For example, suppose you send out 100 resumes. There are only two things that can happen: you get a response (success) or you don't get a response (failure). This is very similar to the tossing of the coin except the chance of success differs from the chance of tossing a head. In addition to the binomial distribution, statisticians have also established formulas for computing the probability of outcomes from different experiments. This chapter deals with these types of distributions.

Chapter Review

This chapter introduces several important discrete distributions and shows how these distributions can be used to solve a wide variety of problems in business and economics.

1. A ***random variable*** is a variable that can take on a certain number of numerical outcomes. Random variables can be either ***discrete*** or ***continuous***. A discrete random variable is one that can take on only a limited number of values. A continuous random variable is one that can take on any possible value within an interval. This chapter deals

with discrete random variables, while Chapters 7 and 8 deal with continuous random variables.

2. The ***binomial distribution*** allows us to calculate the probability of x successes out of n trials when the probability, p, of success in each experiment remains the same for all experiments. Examples of binomial distributions include the probability of receiving 3 heads in 10 flips of a coin or the probability of rolling 20 sixes in 100 rolls of a die.

3. The ***Poisson distribution*** is similar to the binomial distribution except that the probability of success in each experiment is very small and the number of experiments, n, is large. Very often, the information for establishing a Poisson distribution is given in the form of λ (= np), the average number of occurrences in an interval or within a given time period. For example, what is the probability of 12 people entering a bank in a 2 minute time period, given that the average number of people entering the bank in any 2 minute period is 10?

4. The ***hypergeometric distribution*** is similar to the binomial distribution except that it provides probabilities when values are generated without replacement.

5. ***Joint distributions*** deal with the likelihood of the simultaneous occurrence of two or more events. ***Marginal probabilities*** deal with the chance of Event A occurring regardless of the results of Event B. ***Conditional probabilities*** measure the chance of Event B occurring, given that Event A has already occurred.

6. ***Expected values*** deal with the central tendency of the distribution. ***Variance*** and ***standard deviation*** deal with the dispersion of the distribution around its expected value. ***Covariance*** and ***correlation*** measure the degree of association between two random variables.

Useful Formulas

Binomial distribution:

$$P_r(x \text{ suc} \mid n \text{ trials}) = \frac{n!}{x!\,(n-x)!}\, p^x\,(1-p)^{n-x}$$

$$\mu_x = np$$

$$\sigma_x^2 = np(1-p)$$

Poisson distribution:

$$P_r(X = x) = e^{-\lambda}\frac{\lambda^x}{x!}$$

$$\mu_x = \lambda$$

$$\sigma_x^2 = \lambda$$

Hypergeometric distribution:

$$P_r(x|n \text{ trials}) = \frac{C_x^h C_{n-x}^{N-h}}{C_n^N}$$

$$\mu_x = n\left[\frac{h}{N}\right]$$

$$\sigma_x^2 = \left(\frac{N-n}{N-1}\right)\left[n\left(\frac{h}{N}\right)\left(1-\frac{h}{N}\right)\right]$$

Expected value for a discrete random variable:

$$\mu = E(X) = \sum_{i=1}^{N} x_i P(x_i)$$

Variance for a discrete random variable:

$$\sigma^2 = E[x_i - \mu]^2 = \sum_{i=1}^{N}(x_i - \mu)^2 P(x_i)$$

Covariance for two discrete random variables:

$$\mathrm{Cov}(X,Y) \equiv \sigma_{X,Y} = \sum (x_i - \mu_x)(y_i - \mu_y)P_i$$

Correlation coefficient for two discrete random variables:

$$\rho_{X,Y} = \frac{\sigma_{X,Y}}{\sigma_X \sigma_Y}$$

Conditional probability:

$$P(x|y) = \frac{P(x,y)}{P(y)}$$

Marginal probability of x_i:

$$P(x_i) = \sum_{j=1}^{m} P(x_i, y_i)$$

Marginal probability of y_i:

$$P(y_i) = \sum_{i=1}^{n} P(x_i, y_i)$$

Example Problems

Example 1 **Binomial Distribution**

The Kansas City Royals and the Montreal Expos enter the World Series. The Royals are considered to have a 70% chance of winning in each game. The Expos have only a 30% chance of winning. The team that wins 4 out of 7 games will win the series. What is the probability that the Expos will win the championship?

Solution: We can use the binomial distribution to solve this problem with the probability of success equal to .30; x = 4 and n = 7.

$$P_r(x = 4|\ n = 7, p = .30) = {}_7C_4\, .3^4\, (1 - .3)^{7-4}$$
$$= .0972$$

Example 2 **Binomial Distribution - Cumulative Probability**

Using the information given in Example 1, find the probability that the Expos will win 3 games or fewer.

Solution: The probability that the Expos will win 3 games or fewer is simply one minus the probability that the Expos will win the series. Because the Expos can either win the series, ($x = 4$), or lose the series, ($x \leq 3$),

$$P_r(x = 4) + P_r(x \leq 3) = 1.$$

Since $P_r(x = 4) = 0.0972$

$$P_r(x \leq 3) = 1 - P_r(x = 4) = 0.9028$$

Example 3 **Poisson Distribution**

Suppose the average number of customers at a gas station in any 30 minute period is equal to 8. What is the probability that there will be exactly 6 customers in any 30 minute period?

Solution: Because this question uses a time dimension, the Poisson distribution is appropriate.

$$P_r(x = 6) = [e^{-8}\ 8^6]/6!$$
$$= .122$$

Example 4 **Poisson Distribution - Cumulative Probability**

Using the information from Example 3, find the probability that there are more than 5 customers in any 30 minute period.

Solution: The probability that there are more than 5 customers is the complement of 5 customers or fewer. Therefore,

$$P_r(x > 5) = 1 - P_r(x \leq 5)$$
$$= 1 - [P_r(x = 0) + P_r(x = 1) + \ldots + P_r(x = 5)]$$
$$= 1 - [.0003 + .0029 + .0107 + .0286 + .0572 + .0916]$$
$$= .806 \text{ or } 80.6\%$$

Example 5 **Mean and Variance for Discrete Probability Distribution**

Write out the probability distribution for the sum of rolling two dice. Compute the mean and variance for the distribution.

Solution:

Sum S_i	Frequency	Probability P_i	S_iP_i
2	1	1/36	2/36
3	2	1/18	3/18
4	3	1/12	4/12
5	4	1/9	5/9
6	5	5/36	30/36
7	6	1/6	7/6
8	5	5/36	40/36
9	4	1/9	9/9
10	3	1/12	10/12
11	2	1/18	11/18
12	1	1/36	12/36
			252/36
Total	36		

Mean $= \mu_s = \Sigma\, P_i\, S_i$

$= 252/36 = 7$

Var$= \sigma^2 = \Sigma\, P_i\,(S_i - \mu_s)^2$

$= 38.69$

Example 6 **Mean and Variance for Binomial Distribution**

Suppose the probability of receiving a response when you send out a resume is .12. If you send out 55 resumes, find the mean and variance for the distribution.

Solution: $\mu = np = .12 \times 55 = 6.6$

$\sigma^2 = np(1-p) = 55 \times .12 \times (1-.12) = 5.808$

Example 7 **Mean and Variance for Poisson Distribution**

Suppose an average of five orders are placed every 30 minutes at a mail order computer store. Find the mean and variance of the distribution.

Solution: $\mu = \lambda = 5$

$\sigma^2 = \lambda = 5$

Example 8 **Hypergeometric Distribution**

Suppose we have a bag with 5 balls, 3 red and 2 white. If you draw 3 balls out of the bag without replacement, what is the probability that you receive 2 red balls and a white ball? What is the probability if there is replacement?

Solution: Without replacement, we have the hypergeometric distribution.

$$\frac{C_x^h \, C_{n-x}^{N-h}}{C_x^N} = \frac{C_2^3 \, C_{3-2}^{5-3}}{C_3^5} = \frac{\frac{(3)(2)(1)}{(2)(1)(1)} \, \frac{(2)(1)}{(1)(1)}}{\frac{(5)(4)(3)(2)(1)}{(3)(2)(1)(2)(1)}} = .6$$

With replacement, we have the binomial distribution with a 3/5 chance of drawing a red ball. So if we draw 3 balls, the probability of getting 2 red ones is

$$P_r(x=2) = \binom{3}{2}\left(\frac{3}{5}\right)^2\left(1-\frac{3}{5}\right)^{3-2} = .432$$

Example 9 **Expected Value**

A fire insurance policy will pay you $100,000 if your house burns down. If there's a .0005 chance that the house will burn down, what's the expected value of the policy?

Solution: Assume X = the value of the policy, then X will equal 0 or $100,000 depending on whether the house burns down or not.

X	P	PX
0	0.9995	0
100,000	0.0005	50

$$E(X) = \Sigma\, PX = 0(0.9995) + 100{,}000(0.0005) = 50$$

Example 10 **Poisson Approximation to Binomial Distribution**

In the production process, some defective parts are unavoidable. A statistician monitors the process and finds a defect rate of .0001. If 200 parts are randomly selected, what is the probability that there will be more than 2 defective items?

Solution: This is a binomial distribution question because we are interested in the probability of obtaining more than 2 defective items from the 200 items examined. However, using the binomial distribution formula will be very complicated. Fortunately, we can approximate the binomial distribution using the Poisson distribution. For n = 200 and p = .001, $\lambda = np = .2$.

$$P_r(x > 2) = 1 - P_r(x \le 2) = 1 - P_r(x = 0) - P_r(x = 1) - P_r(x = 2)$$

$$= 1 - \frac{e^{-.2}.2^0}{0!} - \frac{e^{-.2}.2^1}{1!} - \frac{e^{-.2}.2^2}{2!} = .0012$$

Example 11 **Expected Value and Standard Deviation**

The following table summarizes the returns of Stocks A and B.

Economy	r_A	r_B	P_r	$r_A \times r_B$
High Growth	10%	20%	40%	2.0%
Recession	8%	4%	60%	.32%

There's a 40% chance of a recession and a 60% chance of high growth next year. Compute the expected rate of return for stocks A and B. Compute the standard deviation of the returns for stocks A and B.

Solution: $E(r_A) = 10\% \times 40\% + 8\% \times 60\% = 8.8\%$

$E(r_B) = 20\% \times 40\% + 4\% \times 60\% = 10.4\%$

$$SD(r_A) = [(10\% - 8.8\%)^2 \times 40\% + (8\% - 8.8\%)^2 \times 60\%]^{1/2} = .0098$$

$$SD(r_B) = [(20\% - 10.4\%)^2 \times 40\% + (4\% - 10.4\%)^2 \times 60\%]^{1/2} = .2021$$

Example 12 **Covariance and Correlation**

Use the information given in Example 11 to find the covariance and correlation between stocks A and B.

Solution:

$$\begin{aligned} Cov(r_A\, r_B) &= E[r_A - E(r_A)]\, [r_B - E(r_B)] \\ &= E(r_A\, r_B) - E(r_A)E(r_B) \\ &= (2\% \times 40\% + .32\% \times 60\%) - 8.8\% \times 10.4\% \\ &= .000768 \end{aligned}$$

$$\begin{aligned} Corr(r_A\, r_B) &= Cov(r_A, r_B)/[SD(r_A)\, SD(r_B)] \\ &= .000768/[(.0098)(.2021) \\ &= .3878 \end{aligned}$$

Example 13 **Portfolio Risk and Return**

In finance, we are interested in reducing the risk (measured by standard deviation) of investments. One way to do this is to diversify our portfolio by holding more than one stock. Using the information from Examples 12 and 13, let's examine the expected value and standard deviation for a portfolio consisting of 50% in stock A and 50% in stock B.

Solution:

$$\begin{aligned} E(r_p) &= E(r_A) \times 50\% + E(r_B) \times 50\% \\ &= 8.8\% \times 50\% + 10.4\% \times 50\% \\ &= 9.6\% \end{aligned}$$

$$\begin{aligned}\text{Var}(r_P) &= 50\%^2\ \text{Var}(r_A) + 50\%^2\ \text{Var}(r_B) + 2\ 50\%(1 - 50\%)\ \text{Cov}(r_A\ r_B) \\ &= 50\%^2\ [.000096 + .04084 + 2(.000768)] \\ &= .011\end{aligned}$$

$$\text{SD}(r_P) = 1 = .103$$

Supplementary Exercises

Multiple Choice

1. A Bernoulli process
   a. gives the number of successes in *n* trials.
   b. consists of one trial with only two possible outcomes.
   c. is the expected value of a binomial distribution.
   d. is a continuous distribution.
   e. can be used for more than two possible outcomes.

2. The hypergeometric distribution is
   a. a continuous distribution.
   b. generated with a Bernoulli process with replacement.
   c. generated with a Bernoulli process without replacement.
   d. the same as the binomial distribution.
   e. the expected value of the binomial distribution.

3. A discrete random variable
   a. can take on an infinite number of values.
   b. can take on a finite and countable number of values.
   c. is another name for a probability mass function.
   d. is always generated from a Bernoulli process.
   e. always has a mean of 0 and a variance of 1.

4. The cumulative binomial function gives us the probability that we will have
   a. more than x successes in n trials.
   b. at least x successes in n trials.
   c. exactly x successes in n trials.
   d. exactly n successes in n trials.
   e. no successes in n trials.

5. The difference between the binomial distribution and the hypergeometric distribution is that
   a. the binomial distribution assumes that the probability of success is constant, whereas the hypergeometric distribution assumes that it changes with each experiment.
   b. the binomial distribution assumes that the probability of success changes, whereas thev hypergeometric distribution assumes that it remains constant.
   c. the hypergeometric distribution is a continuous distribution, whereas the binomial distribution is a discrete distribution.
   d. the hypergeometric distribution is a discrete distribution, whereas the binomial distribution is a continuous distribution.
   e. there is no difference between the two distributions.

6. A continuous random variable is one that can take on
   a. only a countable number of values.
   b. an infinite number of values.
   c. both a countable and an infinite number of values.
   d. only integer values.
   e. only values that are negative.

7. The Poisson distribution
   a. can be approximated by the binomial distribution.
   b. allows us to compute the probability of x successes in n trials.
   c. allows us to compute the probability of x successes in a given time interval.
   d. is a continuous distribution
   e. is another name for a Bernoulli process.

8. In the Poisson distribution both the mean and variance of the distribution are
   a. the number of successes, x, in n trials.
   b. the number of trials, n.
   c. the number of successes, x.
   d. equal to the average amount of success in a given time interval.
   e. equal to x! times the number of trials, n.

9. Suppose the New York Yankees are playing the Los Angeles Dodgers in the World Series (winner is the team that wins 4 out of 7 games). The probability that the Dodgers win 3 games or less
   a. can be determined by the binomial distribution.
   b. is $1 - P_r$(Dodgers win 4 games).
   c. is equal to .5.
   d. can be determined by using a Bernoulli process.
   e. is $1 - P_r$(Yankees win 4 games).

10. The cumulative probability, $F(y)$ is
    a. $P_r(x \geq y)$
    b. $P_r(x = y)$
    c. $P_r(x \leq y)$
    d. always equal to 1.
    e. always equal to 0.

11. The properties of the probability function for discrete random variables are
    a. $P_r(x_i) > 0$ for all I
    b. $P_r(x_i) \geq 0$ for all I
    c. $\Sigma\, P_r(x_i) = 1$
    d. both a and c.
    e. both b and c.

12. The expected value of a discrete random variable, x, can be computed as
    a. $\Sigma\, x_i\, P_r(x_i)$
    b. $\Sigma\, x\, P_r(x_i)$
    c. $\Sigma\, [x_i - E(x_i)]^2\, P_r(x_i)$
    d. $\Sigma\, x_i$
    e. $\Sigma\, x$

13. The variance of a discrete random variable, x, can be computed as
    a. $\Sigma\, x_i\, P_r(x_i)$
    b. $\Sigma\, x\, P_r(x_i)$
    c. $\Sigma\, [x_i - E(x_i)]^2\, P_r(x_i)$
    d. $\Sigma\, x_i$
    e. $\Sigma\, x$

14. The covariance between two random variables, x and y can be computed as
    a. $E[(x - E(x)]\, [y - E(y)]$
    b. $\Sigma\, [y_i - E(y_i)]^2\, P_r(y_i)$
    c. $\Sigma\, [x_i - E(x_i)]^2\, P_r(x_i)$
    d. $\Sigma\, x_i\, P_r(y_i)$
    e. $\Sigma\, x\, P_r(y_i)$

15. The coefficient of correlation between x and y will
    a. always be greater than 0.
    b. always be greater than 1.
    c. be between 0 and 1.
    d. be between –1 and 1.
    e. always be negative.

True/False (If False, explain why)

1. The binomial distribution and Poisson distribution are completely unrelated to each other.
2. When we use a binomial distribution to study the opinion of a population, we are assuming that the population is large.
3. The expected value of a random variable is the value we expect the random variable to realize.
4. When we are sampling with replacement, we should use the hypergeometric distribution.
5. When the population is very large and sampling is done without replacement, the binomial and hypergeometric distributions will give similar results.

6. The only way to determine the probability of tossing exactly 4 heads in 10 flips of a coin is to use the binomial distribution.
7. If the covariance between x and y is 10 and the covariance between c and d is 25, c and d are 2.5 times more strongly correlated than x and y.
8. If the correlation coefficient between x and y is 1, then when variable x goes up we expect variable y to go down.
9. The variance of a binomial distribution with probability of success, p, is equal to np.
10. A call option can be evaluated by using the binomial distribution.

Questions and Problems

1. Suppose a coin is tossed 15 times. What is the probability of tossing heads 3 times or less?
2. A bag contains ten numbers from 1 to 10. If you draw four numbers out of the bag with replacement, what is the probability that all four will be even numbers?
3. Redo Exercise 2 assuming that the four numbers are drawn without replacement.
4. Suppose a grocery store averages 6 customers per 5 minute period. Find the probability that the store will have fewer than 3 customers in any given 5 minute period.
5. You are given the following information about a stock:

   $S = \$50$, $X = \$55$, $r = .005$, $n = 6$, $u = 1.05$, $d = .90$

   Compute the value of the call option.
6. Suppose a coin has a 75% chance of turning up heads and a 25% chance of turning up tails. Find the mean and variance for the number of heads tossed in 1,000 flips of the coin.
7. Suppose you roll a die five times. What is the probability of receiving exactly one roll of six?
8. Suppose you play a game where a coin is tossed. If the toss is a head you receive $100; if the toss is a tail, you receive $125. Compute the expected value and standard deviation for this game.

9. You are given the following rates of return for two stocks in two different climates. Stock B is the stock for a beach resort; Stock P is the stock for an indoor swimming pool. The possible climates and stock returns are given in the following table

Weather	r_B	r_P	P_r	$r_B\ r_P$
Sunny	25%	–5%	.30	–1.25%
Rainy	–5%	25%	.70	–1.25%

Compute the expected value and standard deviation for each stock.

10. Use the information given in Exercise 9 to compute the covariance and correlation coefficient between Stocks B and P.

11. Use the information given in Exercise 9 and your results from Exercise 10 to find the expected value and standard deviation for a portfolio consisting 60% of Stock B and 40% of Stock P.

Answers to Supplementary Exercises

Multiple Choice

1. a	6. b	11. e
2. d	7. c	12. a
3. b	8. d	13. c
4. d	9. b	14. a
5. a	10. c	15. d

True/False

1. False. The Poisson distribution is a special case of the binomial distribution when a binomially distributed random variable is generated from a large number of experiments and the probability of a success in each experiment is very small. In this case, it is easier to use the Poisson distribution to approximate the binomial distribution.
2. True. The population has to be large so that the probability of a certain opinion is not changed from one experiment to another. Otherwise, the hypergeometric distribution is more appropriate.

3. False. The expected value of a random variable should be interpreted as the theoretical average of the random variable and should not be interpreted "literally" as what we expect the random variable to realize.
4. False. When we are sampling with replacement, we should use the binomial distribution. The hypergeometric distribution is used when we sample without replacement.
5. True
6. False. We do not need to use the binomial distribution to determine the probability of tossing exactly 4 heads in 10 flips of the coin, we use the binomial distribution because it makes the computation much easier.
7. False. The covariance between two different pairs of variables cannot be used to compare the degree of correlation.
8. False. When the correlation coefficient between x and y is 1, then when x goes up, we know y will also go up.
9. False. The variance is equal to $np(1 - p)$. The mean is equal to np.
10. True

Questions and Problems

1. .0176
2. .0625
3. .0238
4. .1512
5. \$2.08
6. $\mu = 750$ $\sigma^2 = 187.5$
7. .402
8. $\mu = \$112.50$ $\sigma^2 = 156.25$
9. $E(r_B) = 4$ Standard Deviation $(r_B) = 27.61$

   $E(r_P) = 16$ Standard Deviation $(r_P) = 13.75$
10. Cov $(r_B\ r_P) = -65.25$

    Correlation coefficient $(r_B\ r_P) = -.172$
11. Variance (portfolio) = 273.34

    = 12.88

CHAPTER 7

THE NORMAL AND LOGNORMAL DISTRIBUTIONS

Chapter Intuition

This chapter deals with the most important distribution in statistics, the normal distribution. The ***normal distribution***, which is represented by a bell shaped curve, is useful for conducting many kinds of analyses. It is completely described by two parameters, its mean, μ, and its variance, σ^2. The normal distribution is a probability distribution, which means that the area under the curve is equal to one. The probability that a random variable X takes a value lower than, say, 2, is the size of the area under the normal distribution curve to the left of 2. This area can be found by checking a standard normal distribution table.

The normal distribution is especially important because it can be used to approximate many different types of distributions, including discrete distributions. One important theorem that is used with the normal distribution is the ***central limit theorem***. The central limit theorem says that under certain conditions, the distribution of the sample mean follows a normal distribution, regardless of the distribution of the data. This is very important because many of the analyses we conduct in statistics use the sample mean. The central limit theorem will be discussed in Chapter 8 of the text.

The second type of distribution discussed in this chapter is the ***lognormal distribution***. A random variable X is said to have a lognormal distribution if log(X) follows a normal distribution. One of the properties of the lognormal distribution is that it is only valid for values of X that are positive. This can be useful for describing variables that cannot take on negative values, such as stock prices.

Chapter Review

1. The ***normal distribution*** is the most important and widely used distribution in statistics. It has a bell-shaped curve. The normal distribution is symmetrical around the mean, and is completely described by its mean, μ, and variance, σ^2. Below we show how the distribution changes as the variance changes. The smaller the variance, the less spread out the distribution is around the mean.

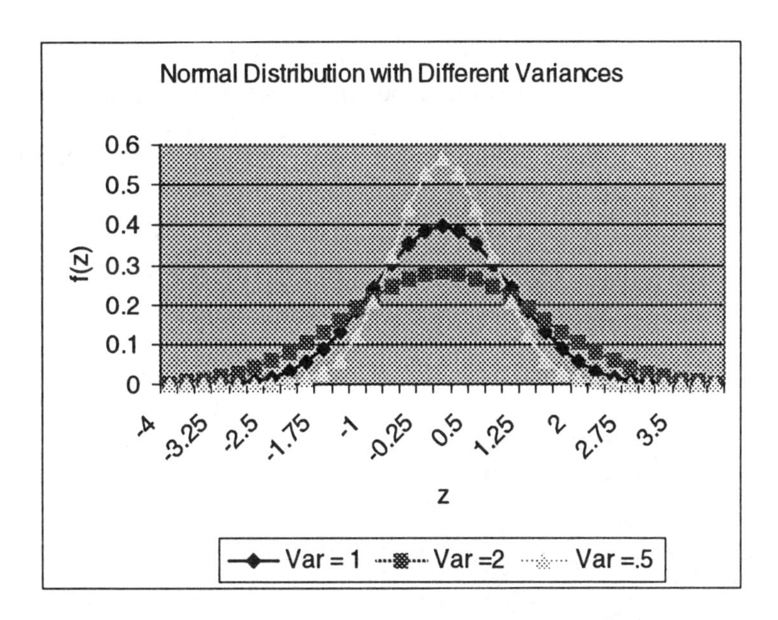

The normal distribution is also described by its mean. Below we show the case of two normal distributions, one with a mean of 0 and the other with a mean of 2. Notice how the mean determines the center of the distribution.

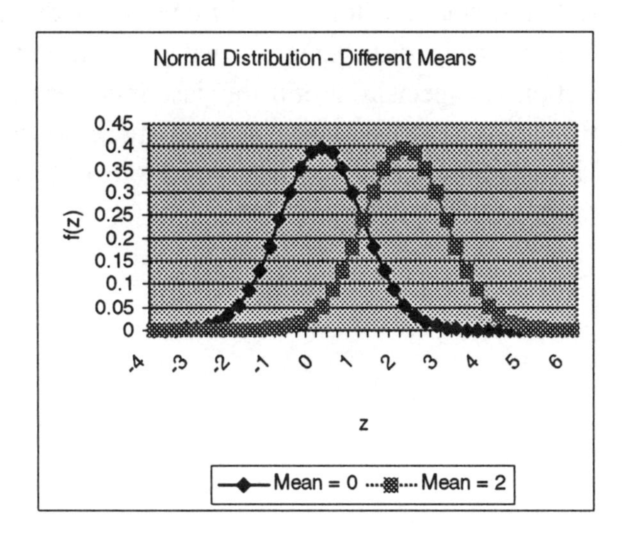

2. The standard normal variable, z, refers to a normal distribution with a mean of zero and a standard deviation of one. The standard normal distribution table in the text gives us the probability that z will have a value between 0 and a given critical value. To use the

normal distribution table, we need to transform a normal variable X into a standard normal variable by $z = (X - \mu)/\sigma$, where μ is the mean and σ is the standard deviation.

3. Because it is a very cumbersome task to compute cumulative probabilities using discrete distributions such as the binomial and Poisson distributions, we often use the normal distribution to approximate these discrete distributions. The normal distribution gives us a fairly accurate approximation when the sample size is large. Below is a graph of the cumulative normal distribution function.

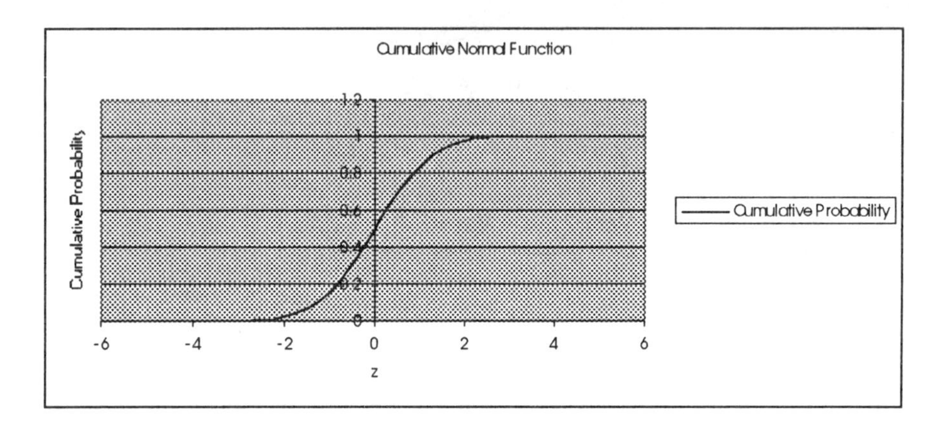

4. The ***lognormal distribution*** represents a simple transformation of a normally distributed variable Y. Let $X = e^Y$. If Y is normally distributed, X will have a lognormal distribution. The lognormal distribution is especially useful for describing variables in business and economics that cannot take on negative values, such as stock prices. The graph below shows the lognormal distribution with a standard deviation of 1 and two different means.

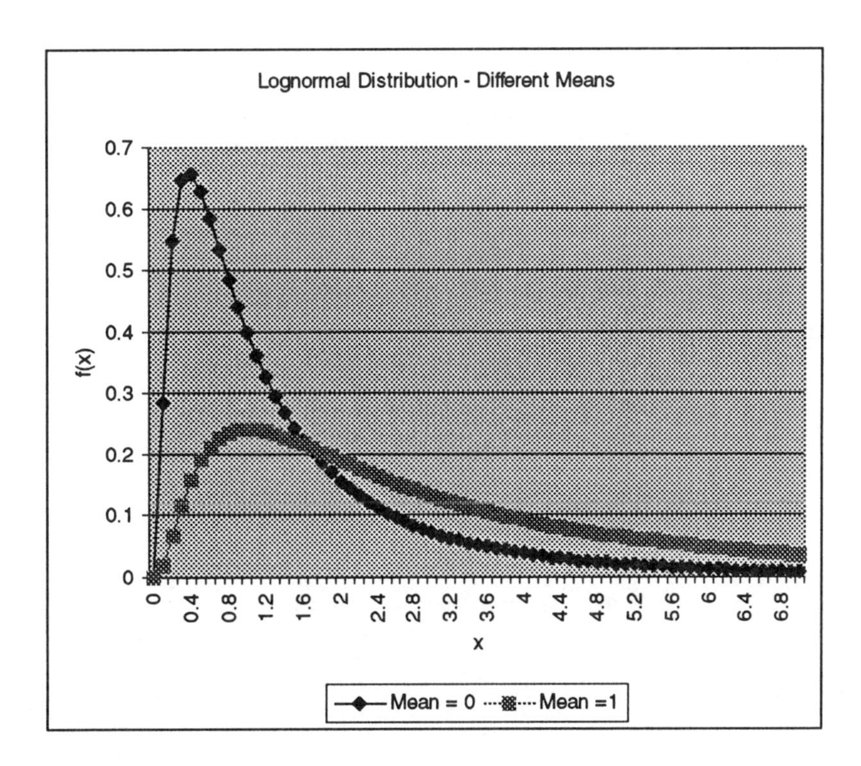

5. One application of the normal distribution in finance is the Black-Scholes Option Pricing Model. The Black-Scholes formula uses cumulative normal probabilities to compute the value of a call option.

Useful Formulas

Standard normal variable:

$$z = \frac{x - \mu_x}{\sigma_x}$$

Normal approximation of binomial:

$$z = \frac{x - np}{\sqrt{np(1 - np)}}$$

Normal approximation of Poisson:

$$z = \frac{x - \lambda}{\sqrt{\lambda}}$$

Black-Scholes option pricing formula:

$$P_o = P_S N(d_1) - Ee^{-rt} N(d_2)$$

$$d_2 = d_1 - \sigma\sqrt{t}$$

$$d_1 = \frac{\ln\left(\frac{P_S}{E}\right) + \left(r + \frac{\sigma^2}{2}\right)t}{\sigma\sqrt{t}}$$

Probability that *x* lies between *a* and *b* (discrete random variable):

$$P(a \leq x \leq b) = \sum_{x=a}^{b} f(x)$$

Probability that *x* lies between *a* and *b* (continuous random variable):

$$P(a < x < b) = \int_a^b f(x)dx$$

Lognormal distribution:

$$X = e^Y \qquad\qquad Y \sim N(\mu, \sigma^2)$$

$$\mu_X = e^{\mu+\sigma^2/2}$$

$$\sigma_X^2 = e^{2\mu+\sigma^2(e^{\sigma^2}-1)}$$

Example Problems

Example 1 Area Under the Normal Distribution

Calculate the area under the standard normal curve between the following values.

a. $z_1 = 0$ and $z_2 = 1.0$
b. $z_1 = -3.5$ and $z_2 = -1$
c. $z_1 = -1.0$ and $z_2 = 1.0$
d. $z_1 = -3.0$ and $z_2 = 0$

Solution: The areas can be found by looking at the standard normal distribution table. Because the normal distribution is symmetric with mean equal to zero, the area under the curve below zero equals the area under the curve above zero.

a. $P_r(0 < z < 1.0) = .3413$
b. $P_r(-3.5 < z < -1) = .1576$
c. $P_r(-1 < z < 1) = .6826$
d. $P_r(-3 < z < 0) = .4987$

Example 2 Standardizing Normal Random Variables

Calculate the area under the normal curve between the following values given that the mean is 3 and the variance is 2.

a. $x_1 = 7$ and $x_2 = 8.0$
b. $x_1 = -3.5$ and $x_2 = -1$
c. $x_1 = -1.0$ and $x_2 = 1.0$
d. $x_1 = -4.0$ and $x_2 = 2.0$

Solution: Because these variables are not standard normal, they must be converted to standard normal variables, using the following formula:

$z = (x - \mu)/\sigma$

Once the variables have been standardized, the procedure is the same as for Example 1.

a. $z_1 = [7 - 3]/\sqrt{2}$
$= 2.83$

$z_2 = [8 - 3]/\sqrt{2}$
$= 3.54$

$P_r(2.83 < z < 3.54) = .0021$

b. $z_1 = [-3.5 - 3]/\sqrt{2}$
$= -4.6$

$z_2 = [-1 - 3]/\sqrt{2}$
$= -2.83$

$P_r(-4.6 < z < -2.83) = .023$

c. $z_1 = [-1 - 3]/\sqrt{2}$
$= -2.83$

$z_2 = [1 - 3]/\sqrt{2}$
$= 1.41$

$P_r(-2.83 < z < 1.41) = .9169$

d. $z_1 = [-4 - 3]/\sqrt{2}$
$= -4.95$

$z_2 = [2 - 3]/\sqrt{2}$
$= .707$

$P_r(-4.95 < z < .707) = .7580$

Example 3 **Normal Approximation to Binomial Distribution**

Using the normal approximation to the binomial distribution with $n = 100$ and $p = .2$.

a. What is the probability that a value from the distribution will be greater than 25?

b. What is the probability that a value from the distribution will be less than 10?

Solution: The mean for the binomial distribution is np, and the variance is np(1 – p). In order to use the normal approximation to the binomial distribution, we need to convert the random variables to standard normal values.

a. $z = [25 - (100)(.2)]/[(100)(.2)(1 - .2)]^{1/2}$
$= 1.25$

$P_r(z > 1.25) = .0885$ or 8.85%

b. $z = [10 - (100)(.2)]/[(100)(.2)(1 - .2)]^{1/2}$
$= -2.5$

$P_r(z < -2.5) = .0062$ or .62%

Example 4 **Normal Approximation to Poisson Distribution**

Using the normal approximation to the Poisson distribution with $\lambda = 15$:

a. What is the probability that a value from the distribution will be greater than 10?

b. What is the probability that a value from the distribution will be between 20 and 30?

Solution: Again, we need to find the standard normal values for the x variables. The Poisson distribution has mean and variance of λ.

a. $z = [10 - 15]/\sqrt{15}$
$= -1.29$

$P_r(z > -1.29) = .9015$ or 90.15%

b. $z_1 = [20 - 15]/\sqrt{15}$
$= 1.29$

$z_2 = [30 - 15]/\sqrt{15}$
$= 3.87$

$P_r(1.29 < z < 3.87) = .0975$ or approximately 9.75%

Example 5 **Cumulative Probabilities**

A quality control manager has computed that the life of light bulbs follows a normal distribution with a mean of 300 and a standard deviation of 15. Compute the probability that the number of defective light bulbs in a case is

a. greater than 320.

b. less than 290.

c. between 320 and 330.

Solution: Before we can use the normal distribution table, we must standardize the variables.

a. $z = [320 - 300]/15$
$= 1.33$

$P_r(z > 1.33) = .0918$ or 9.18%

b. $z = [290 - 300]/15$
$= -.667$

$P_r(z < -.667) = .2514$ or 25.14%

c. $z_1 = [320 - 300]/15$
$= 1.33$

$z_2 = [330 - 300]/15$
$= 2$

$P_r(1.33 < z < 2) =$ .064 or 6.4%

Example 6 **Area Under the Normal Distribution**

The manager in a local convenience store finds that the number of gallons of milk sold in a week follows a normal distribution with a mean of 200 and a standard deviation of 40. If the manager wants to make sure that there is a sufficient amount of milk in the store at the beginning of the week so there is only a 5% chance that the store will run out of milk, what is the minimum amount of milk she should purchase at the beginning of the week?

Solution: A graph can help illustrate the answer. Assume that the amount of milk demanded in a week is x, and that we know that x follows a normal distribution with a mean of 200 and a standard deviation of 40. We should find a minimum amount of milk, x_{min} so that there is only a 5% chance that the milk demanded, x, will be greater than this minimum amount, as shown in the following graph.

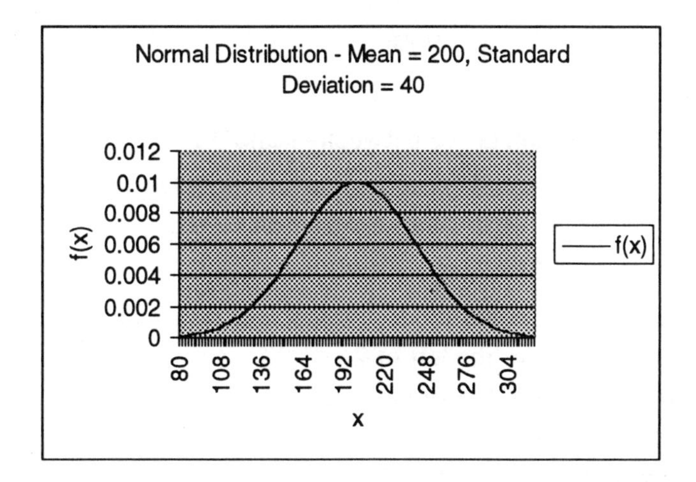

$P_r(x > x_{min}) = 5\%$

$P_r[(x - 200)/40 > (x_{min} - 200)/40] = 5\%$

$P_r[1.64 > (x_{min} - 200)/40] = 5\%$

$1.64 = (x_{min} - 200)/40$

Solving for x_{min} yields, $x_{min} = 265.6$

Supplementary Exercises

Multiple Choice

1. The normal distribution is
   a. a discrete distribution.
   b. approximated by the binomial distribution.
   c. a continuous distribution completely described by its mean and variance.
   d. approximated by the Poisson distribution.
   e. is positively skewed.

2. $P_r(z < -1.64) =$
   a. 5%
   b. 95%
   c. –5%
   d. –95%
   e. 90%

3. The area under the normal distribution curve between -1 and 3 is
   a. $P_r[x < 3] - P_r[x < -1]$
   b. $P_r[x > 3] - P_r[x < -1]$
   c. $P_r[x < 3] + P_r[x < -1]$
   d. $P_r[x > 3] + P_r[x < -1]$
   e. $P_r[x = 3] + P_r[x = -1]$

4. The probability that a normally distributed random variable, x, is greater than 2 is the
   a. area under the normal distribution to the left of 2.
   b. area under the normal distribution to the right of 2.
   c. height of the normal distribution at 2.
   d. area under the entire normal distribution curve.
   e. area under the entire normal distribution curve divided by 2.

5. Suppose we have two normally distributed random variables, x and y. Variable x has a mean of 100 and a standard deviation of 20, whereas y has a mean of 100 and a standard deviation of 40. If the two distributions are plotted on the same graph, the
   a. distribution of y will be to the right of the distribution of x.
   b. distribution of y will be to the left of the distribution of x.
   c. two distributions will have the same center; however, the distribution of y will be flatter.
   d. two distributions will have the same center, but the distribution of y will be steeper.
   e. distribution of y will be skewed to the left and the distribution of x will skewed to the right.

6. Suppose we have two normally distributed random variables, x and y. Variable x has a mean of 100 and a standard deviation of 20, whereas y has a mean of 200 and a standard deviation of 20. If the two distributions are plotted on the same graph, the
   a. distribution of y will be to the right of the distribution of x.
   b. distribution of y will be to the left of the distribution of x.
   c. two distributions will have the same center, but the distribution of y will be flatter.
   d. two distributions will have the same center, but the distribution of y will be steeper.
   e. distribution of y will be skewed to the left and the distribution of x will be skewed to the right.

7. A continuous random variable can
   a. take on only a finite and countable number of values.
   b. take on an infinite number of values.
   c. be approximated by a discrete random variable.
   d. only take on integer values.
   e. never be negative.

8. The Black-Scholes model
   a. is a formula for analyzing stock prices.
   b. explains why the normal distribution is the most important distribution in statistics.
   c. uses the normal distribution to value options.
   d. is used to approximate the binomial distribution.
   e. is used to approximate the Poisson distribution.

9. The lognormal distribution
   a. is valid for nonpositive numbers.
   b. is valid for nonnegative numbers.
   c. can take on any value.
   d. is rarely ever used in finance and accounting.
   e. is a discrete distribution.

10. The lognormal distribution is especially useful for describing
    a. stock prices.
    b. earnings per share.
    c. stock returns.
    d. price/earnings ratios.
    e. earnings growth.

11. Approximately 95% of the area under the normal distribution lies
    a. below the mean.
    b. above the mean.
    c. 1 standard deviation away from the mean.
    d. 2 standard deviations away from the mean.
    e. 3 standard deviations away from the mean.

12. The cumulative normal probability at the mean of the distribution is
    a. 1.
    b. 0.
    c. .5
    d. .25
    e. .75

13. A normally distributed random variable can be standardized by
    a. adding the mean and dividing by the standard deviation.
    b. subtracting the mean and multiplying by the standard deviation.
    c. subtracting the mean and dividing by the standard deviation.
    d. converting to the lognormal distribution.
    e. using the Black-Scholes model.

14. $P_r(z > -1.645) =$
    a. 5%
    b. 95%
    c. –5%
    d. –95%
    e. 90%

15. $P_r(z < 1.96) =$
    a. 2.5%
    b. 97.5%
    c. –2.5%
    d. –97.5%
    e. 95%

16. $P_r(z < -1.96) =$
    a. 2.5%
    b. 97.5%
    c. –2.5%
    d. –97.5%
    e. 95%

True/False (If false, explain why)

1. The binomial distribution can always be approximated using the normal distribution.
2. The Poisson distribution can be approximated using the normal distribution.
3. The normal distribution can have a negative variance.
4. The normal distribution can be positively or negatively skewed.
5. There are an infinite number of normal distributions.
6. The normal distribution can have a variance equal to 0.
7. The mean, median, and mode are always the same for the normal distribution.
8. The lognormal distribution is better for describing stock prices than the normal distribution, because stock prices can never be negative.
9. The probability density function gives us the probability that the random variable, x, will be less than some given value.
10. The area under the normal distribution curve will depend on the mean and standard deviation.
11. The normal distribution is a discrete distribution.
12. The normal distribution is valid for values of x ranging from $-\infty$ to $+\infty$.
13. The lognormal distribution is valid for values of x ranging from $-\infty$ to $+\infty$.
14. The lognormal distribution is a bimodal distribution.

Questions and Problems

1. Compute the z value for $x = 25$ if $\mu = 20$, and $\sigma = 2$.
2. During final exam week, a college professor has an average of 12 students per hour during her office hours. Use the normal approximation to the Poisson to find the probability that she will have fewer than 10 students in any given hour.
3. Suppose you toss a fair coin 50 times. Use the normal approximation to the binomial distribution to find the probability that you will get at least 28 tails.

4. Calculate the area under the normal curve between the following values given that the mean is 125 and the standard deviation is 20.

   a. $x_1 = 90$ and $x_2 = 105$

   b. $x_1 = 125$ and $x_2 = 155$

5. Use the Black-Scholes option pricing formula to compute the value of a call option given the following information.

   S = \$50 price of stock
   E = \$45 exercise price
   r = .045 risk-free interest rate
   t = .55 years till option expires
   σ = .15 standard deviation in the stock's returns

6. Suppose you work for a company that uses batteries. Every time a truckload of batteries arrives, you examine 400 of the 10,000 batteries delivered. Your company policy says that you should refuse any truckload with more than four bad batteries. Suppose a truck load of batteries arrives that has a total of 200 bad batteries. What is the probability that you accept this shipment? (Hint: What is the probability that you will find 4 bad batteries in the 400 you examine, given that 200 of the 10,000 are bad?)

7. Suppose the number of cups of yogurt sold by a convenience store each week follows a normal distribution with a mean of 600 and a standard deviation of 20. If the store owner wants to make sure that he does not have unsold yogurt in more than 5% of the weeks, how many cups of yogurt should he order?

Answers to Supplementary Exercises

Multiple Choice

1. c
2. c
3. a
4. b
5. c
6. a
7. b
8. c
9. b
10. a
11. d
12. c
13. c
14. b
15. d
16. a

True/False

1. False. The binomial distribution can be approximated by the normal distribution when the sample size is large enough.
2. True
3. False. Variance can never be negative for any distribution.
4. False. The normal distribution is always symmetric.
5. True
6. False. When a distribution has a variance equal to zero, all values from the distribution will be identical.
7. True
8. True
9. False. The cumulative distribution function gives us the probability that the random variable, x, will be less than some given value.
10. False. Because the normal distribution is a probability distribution, the area under the curve will always be equal to 1, regardless of the mean and standard deviation.
11. False. Continuous distribution.
12. True
13. False. Valid only for positive numbers.
14. False. Unimodal distribution.

Questions and Problems

1. $z = 2.5$
2. .281
3. .4052
4. a. .1186

   b. .3944
5. $6.41

6. We know that 200 out of 10,000 batteries are bad. This means that 2% of the batteries are bad. If we sample 400 batteries, the probability of returning the shipment is

$$P_r(x > 4) = P_r\left(\frac{x}{n} > \frac{4}{400}\right)$$

Using the normal distribution to approximate the binomial distribution, we get $P_r(z > 4.52) = 0$

So it is almost impossible for us to return this shipment.

7. $P_r(x < 600) = 5\%$

$$P_r\left(z < \frac{x - 600}{20}\right) = 5\%$$

$$\frac{x - 600}{20} = -1.645$$

Solving for x gives us x = 567.1 or approximately 567 cups of yogurt.

CHAPTER 8

SAMPLING AND SAMPLING DISTRIBUTIONS

Chapter Intuition

Often we are interested in the parameters of a population. For example, a manager may be interested in knowing whether the average life of the light bulbs produced yesterday equals or exceeds 300 hours. One way to conduct this analysis is to use a census; that is, light up all of the light bulbs produced yesterday and determine the average life. This approach is both impractical and costly. An alternative approach is to use a sample of 100 light bulbs to determine the average life of all the light bulbs. Because we are using a sample rather than the whole population, errors can occur when we try to make inferences about the entire population. In our light bulb example, it is possible that the average life of the light bulbs in the population is high (for example, 350 hours); however, when we draw our sample, the sample average is only 220 hours. Using this on the sample mean of 220, we erroneously conclude that the average life of all the light bulbs is low. Sampling distributions allow us to determine the probability that we will make this type of mistake.

Using sampling distributions, we can determine the probability distribution of statistics such as the sample mean. Without sampling distributions, we may use the sample mean to infer the population mean; however, we will not know the probability of reaching the correct conclusion.

This chapter builds on the basic concepts of probability and probability distributions discussed in previous chapters and applies these concepts to sampling. This chapter also forms a building block for inferential statistics, which is discussed in part III of the text.

Chapter Review

1. When analyzing a population, either a ***census*** or a ***sample*** can be used to conduct the analyses. In a census, all members of the ***population*** are examined. In a sample, only a subset of the entire population is examined. The advantage of using a sample over a census is that the costs are greatly reduced because we are dealing with a much smaller set of numbers. The cost of sampling is that a sample is usually less accurate than a census.
2. Once we have collected a sample, we are dealing with information about a population which is not known with certainty. In order to quantify this uncertainty, we use the probability concepts discussed in Chapter 5 and the distribution theory of Chapters 6-8.
3. Often we are interested in the sampling distribution of the mean. The ***central limit theorem*** allows us to find the sampling distribution of the sample mean. The central limit theorem says that if random samples of n observations are taken from any population with mean μ_x and standard deviation σ_x, and if n is large enough, the distribution of the mean will be approximately normal with mean $\mu x = \mu_x$ and standard deviation $\sigma x = \sigma/n$, regardless of the population distribution.
4. The most basic procedure for selecting a sample is ***simple random sampling***. Simple random sampling is a procedure in which every member of the population has an equal probability of being selected.
5. When a sample is used rather than a census, errors may occur. ***Sampling errors*** are errors resulting from the chance selection of the sample. Because these errors occur by the chance selection of the sample, they are ***random errors***. ***Non-sampling errors*** are errors that result from inaccurate measurement of the data or improper selection of the sample. Non-sampling error is a ***systematic error*** because it affects all members of the sample in a similar manner. For example, using a ruler that is too short to measure the height of the basketball team. Unlike random error, systematic error cannot be eliminated by increasing the sample size.

Useful Formulas

Sample mean:

$$\overline{X} = \frac{1}{n}\sum_{i=1}^{n} X_i$$

Sample variance:

$$s_x^2 = \frac{1}{n-1}\sum_{i=1}^{n}\left(X_i - \overline{X}\right)^2$$

Standard deviation of the sample mean:

$$\sigma_{\overline{x}} = \frac{\sigma_x}{\sqrt{n}}$$

Mean for a sample proportion:

$$E(\overline{p}_x) = p$$

Variance of a sample proportion:

$$\sigma^2_{\overline{p}_x} = \frac{p(1-p)}{n}$$

Example Problems

Example 1 **Sample vs. Population**

Suppose a city consists of 16,525 people with 3,721 registered Democrats, 3,622 registered Republicans and 2,202 independents. If you are interested in who the Democrats will nominate for the next mayoral election, which group would constitute the population? Give an example of a sample from this population.

Solution: Since we are interested in who the Democrats will nominate, our population consists of all registered Democrats. A sample would consist of a subset of our 3,721 registered Democrats, for example the first 200 registered Democrats alphabetically would be one way to choose a sample. However, we should be careful when choosing a sample in this manner, because it is not clear that the first 200 people alphabetically will represent the viewpoint of all registered Democrats. This type of problem might occur if a large family all with the last name beginning with A, represents a large proportion of the registered Democrats.

Example 2 **Sampling and Nonsampling Error**

State whether the following represent sampling or nonsampling error:

a. The sample thickness of plywood is measured with an inaccurate ruler.

b. A sample of new home prices in the United States shows a mean of $98,200 when the actual mean new home price is $83,100.

c. A restaurant owner asks the first 25 people leaving a restaurant about how they liked the food. However, by chance, the first 25 people are all teenagers.

Solution:

a. This represents a case of nonsampling error because the error is due to the inaccurate ruler not to the sample selected.

b. This represents sampling error. It is simply by chance that the sample we selected has a mean home price above the actual home price. We could just as easily have chosen homes which would have given us a mean home price below the actual home price.

c. Once again we have sampling error. It is simply by chance that the first 25 people the owner surveys are teenagers.

Example 3 **Distribution of the Sample Mean**

The mean life of radial tires produced by the Sure Grip Tire Company, is 45,000 miles, with a standard deviation of 6,000 miles. Suppose you take a random sample of 15 tires.

a. What is the mean of the sample mean life?

b. What is the variance of the sample mean?

Solution: a. $E(\overline{x}) = 45{,}000$

b. $$\begin{aligned} Var(\overline{x}) &= Var(x) / n \\ &= 6{,}000^2 / 15 \\ &= 2{,}400{,}000 \end{aligned}$$

Example 4 **Probabilities of the Sample Mean**

Suppose the time a customer waits at a bank is normally distributed with a mean of 16 minutes and a standard deviation of 5 minutes. If you take a random sample of 5 customers, what is the probability that the average wait time will be at least 12 minutes? What is the mean of the average wait time? What is the standard deviation of the mean waiting time?

Solution: $\mu = 16$ minutes

$$\sigma_{\overline{x}} = \sigma_x / \sqrt{n} = 5 / \sqrt{5} = 2.24$$

$$\begin{aligned} z &= (\overline{x} - \mu) / \sigma_{\overline{x}} \\ &= (12 - 16) / 2.24 = -1.78 \end{aligned}$$

$$P_r = (\overline{x} > 12) = P_r(z > -1.78) = .9625$$

Example 5 **Sample Proportions**

Rah Rah University accepts 60% of all students who apply for admissions. A random sample of 100 applications is taken.

a. What is the probability that more than half the applicants sampled are accepted?

b. What is the probability that the sample acceptance rate is between .50 and .75?

Solution:

$$p = 0.6$$
$$\sigma_p = [.6(1-.6)/100]^{1/2}$$
$$=.049$$

a.

$$P_r(\bar{p} > .5) = P_r(z > (.5-.6)/.049))$$
$$= P_r(z > -2.04)$$
$$=.9793$$

b.

$$P_r = (.5 < \bar{p} < .75)$$
$$= P_r(-2.04 < z < 3.06)$$
$$=.9782$$

Example 6 **Sample Proportion**

From past history, an auto dealer knows that 8% of all customers entering the showroom make a purchase. Suppose that 100 people enter the showroom.

a. What is the mean of the sample proportion of customers making a purchase?

b. What is the variance of the sample proportion?

c. What is the standard deviation of the sample proportion?

d. What is the probability that the sample proportion is between .05 and .10?

Solution:

a. $\mu = .08$

b. $\sigma^2 = [p(1 - p)]/n$
$= [.08(1 - .08)]/100$
$= .000736$

c. $\sigma = (\hat{\sigma}^2) = (.000736)^{1/2} = .0271$

d. $z_1 = (.05 - .08)/.0271$
$= -1.11$

$z_2 = (.10 - .08)/.0271$
$= .738$

$$\begin{aligned} P_r(.05 < p < .10) &= P_r(-1.11 < z < .738) \\ &= P_r(z < .738) - P_r(z < -1.11) \\ &= .7704 - .1335 \\ &= .6369 \text{ or } 63.69\% \end{aligned}$$

Example 7 **Variance of the Sample Mean**

Find the variance of the sample mean or proportion taken from the following populations.

a. Suppose a random sample of 100 students is taken from a student body of 5,000. The students are asked whether they support the new alcohol policy on campus. Assume 3,000 students support the new policy.

b. A random sample of 10 cans of dog food is taken from a case of 24 cans. The mean weight of the dog food is 16 ounces with a standard deviation of .5 ounces.

c. A random sample of 10 cans of dog food is taken from the daily production of an assembly line. The mean is 16 ounces and the standard deviation is .5 ounces.

Solution:

a. $$\frac{p(1-p)}{n} = \frac{0.6(1-0.6)}{100} = 0.0024$$

b. $$\frac{\sigma}{\sqrt{n}}\sqrt{\frac{N-n}{N-1}} = \frac{0.5}{\sqrt{10}}\sqrt{\frac{24-10}{24-1}} = 0.123$$

c. $$\frac{\sigma}{\sqrt{n}} = \frac{0.5}{\sqrt{n}} = 0.158$$

Example 8 **Distribution of the Sample Mean**

In each of the following cases, find the mean and the standard deviation of the sample mean, for a sample of size n from a population with mean μ and standard deviation of σ. Assume the population is large.

a. $n = 25,\ \mu = 6, \sigma = 2$

b. $n = 25,\ \mu = 6, \sigma = 3$

c. $n = 20,\ \mu = 45, \sigma = 21$

d. $n = 40,\ \mu = 120, \sigma = 50$

Solution:

$$\bar{x} = \mu$$

$$\sigma_{\bar{x}} = \sigma_x / \sqrt{n}$$

a. $$\bar{x} = 6$$

$$\sigma_{\bar{x}} = 2/\sqrt{25}$$

$$=.4$$

b. $\bar{x} = 6$

$\sigma_{\bar{x}} = 3 / \sqrt{25}$

$= .6$

c. $\bar{x} = 45$

$\sigma_{\bar{x}} = 21 / \sqrt{20}$

$= 4.70$

d. $\bar{x} = 120$

$\sigma_{\bar{x}} = 50 / \sqrt{40}$

$= 7.90$

Example 9 **Central Limit Theorem**

A sample of 200 cans of soda is taken. If a statistician does not know the original distribution of the soda, can he make a statistical inference of the mean based on the sample? If a sample of 20 cans is taken, what kind of assumption is needed before he conducts any statistical inference? If one wants to estimate the mean and the standard deviation of the sample mean, is the above assumption needed?

Solution: Under certain conditions, the statistical inference based on a large sample can be conducted using the central limit theorem. If the sample is small, we need to know the distribution of the population. However, if we are only interested in estimating the mean and standard deviation, we don't need to make any assumptions about the distribution.

> Example 10 **Distribution of the Sample Mean**
>
> Suppose we have a population of four numbers: 6, 8, 10, and 12. If we randomly draw two numbers without replacement, what is the probability distribution, the mean, and the standard deviation of the sample mean?

Solution: The sample space and sample mean are

Sample Space

(6, 8)	7
(6, 10)	8
(6, 12)	9
(8, 10)	9
(8, 12)	10
(10, 12)	11

The distribution of the sample mean is

$\bar{x}$	$P_r(\bar{x})$	$\bar{x}P_r(\bar{x})$	$[\bar{x}-E(\bar{x})]^2 P_r(\bar{x})$
7	1/6	7/6	4/6
8	1/6	8/6	1/6
9	1/3	9/3	0
10	1/6	10/6	1/6
11	1/6	11/6	4/6
Total		9	10/6

$$E(\bar{x}) = 9$$

$$Var(\bar{x}) = \sigma^2 / n[(N-n)/(N-1)]$$

$$5/2 \times 2/3 = 5/3$$

Example 11 **Distribution of a Sample Proportion**

In a town of 300 residents, 30 percent approve of the new garbage collection policy. If the local newspaper asks 120 residents about their opinion of the new garbage collection policy, what is the probability that more than 30 people will be in favor of it?

Solution: The distribution of is

$$\bar{p} \sim N\left[p, \left(\sqrt{\frac{p(1-p)}{n}}\sqrt{\frac{N-n}{N-1}}\right)^2\right]$$

Since p = 0.3

$$\sqrt{\frac{p(1-p)}{n}}\sqrt{\frac{N-n}{N-1}} = \sqrt{\frac{0.3(1-0.3)}{120}}\sqrt{\frac{300-120}{300-1}}$$

$$= P_r\left(\bar{p} > \frac{30}{120}\right) = P_r\left(\frac{\bar{p}-0.3}{0.0325} > \frac{0.25-0.3}{0.0325}\right)$$

$$= P_r(z > -1.54)$$

$$= 0.9382$$

Note: In this question, the sample is large relative to the population.

Example 12 **Probability**

Suppose you play a game where you roll two dice one time. You win if you roll a sum of 9 or more. What is the probability of winning?

Solution:

Sum	Sample	P_r
2	(1, 1)	1/36
3	(1, 2) (2, 1)	2/36
4	(1, 3) (2, 2) (3, 1)	3/36
5	(1, 4) (2, 3) (3, 2) (4, 1)	4/36
6	(1, 6) (2, 4) (3, 3) (4, 2) (5, 1)	5/36
7	(1, 6) (2, 5) (3, 4) (4, 3) (5, 2) (6, 1)	6/36
8	(2, 6) (3, 5) (4, 4) (5, 3) (6, 2)	5/36
9	(3, 6) (4, 5) (5, 4) (6, 3)	4/36
10	(4, 6) (5, 5) (6,4)	3/36
11	(5, 6) (6, 5)	2/36
12	(6, 6)	1/36

$$\begin{aligned} P_r(x \geq 9) &= P_r(x = 9) + P_r(x = 10) + P_r(x = 11) + P_r(x = 12) \\ &= 4/36 + 3/36 + 2/36 + 1/36 \\ &= 10/36 \end{aligned}$$

Example 13 **Expected Value**

Suppose five chips numbered 2, 4, 6, 8, and 10 are placed in a hat. You are allowed to choose two chips without replacement. The sum of the numbers indicates the dollar payment you receive. How much money should you pay for this game to make it a fair game?

Solution:

Sum of 2 Chips	Two Chips Drawn	P_r
6	(2, 4)	.1
8	(2, 6)	.1
10	(2, 8) (4, 6)	.2
12	(2, 10) (4, 8)	.2
14	(4, 10) (6, 8)	.2
16	(6, 10)	.1
18	(8, 10)	.1

The price you should pay is the expected payoff.

$$\text{Expected Payoff} = \Sigma P_{ri} \times \text{Payoff}_i$$
$$= 6(.1) + 8(.1) + 10(.2) + 12(.2) + 14(.2) + 16(.1) + 18(.1)$$
$$= 12$$

Supplementary Exercises

Multiple Choice

1. A census is conducted by
   a. surveying a subset of the population of interest.
   b. surveying all members of the population of interest.
   c. either surveying all members or a subset of the population of interest.
   d. surveying half of the population.
   e. surveying more than half the population.
2. A sample is conducted by
   a. surveying a subset of the population of interest.
   b. surveying all members of the population of interest.
   c. either surveying all members or a subset of the population of interest.
   d. surveying half of the population.
   e. surveying more than half the population.

3. The advantage of sampling is that it can
   a. reduce costs.
   b. reduce the time spent on data collection.
   c. increase data manageability.
   d. sometimes be more accurate than a census.
   e. all of the above.

4. Sampling errors are
   a. caused by inaccurate measurement.
   b. the result of the chance selection of the sampling units.
   c. of no great concern.
   d. larger for a census than for a sample.
   e. always positive for a census.

5. Nonsampling errors are
   a. caused by inaccurate measurement.
   b. the result of the chance selection of the sampling units.
   c. of no great concern.
   d. always larger for a census than for a sample.
   e. never 0.

6. If μ_x is the population mean, and σ_x^2 is the population variance, then the mean and variance of a sample are equal to
   a. μ_x and σ_x^2
   b. μ_x/n and σ_x^2 / n
   c. μ_x/n and σ_x^2 / n^2
   d. μ_x and σ_x^2 / n
   e. μ_x and $(\sigma_x^2)n$

7. The central limit theorem says that as the sample size, *n*, from a given population gets large enough, the sampling distribution of the mean can be approximated by
   a. the binomial distribution.
   b. the normal distribution.
   c. the Poisson distribution.
   d. a Bernoulli process.
   e. different distributions, depending on the given population.

8. The finite population multiplier is used to
   a. reduce bias in the sample mean.
   b. reduce bias in the sample variance.
   c. make the computation of the variance easier.
   d. make the computation of the mean easier.
   e. reduce sampling error.

9. The mean of a binomially distributed random variable with n experiments and p as the probability of success for each experiment is
   a. np
   b. np(1 – p)
   c. np/[np(1 – p)]
   d. n
   e. p

10. The variance of a binomially distributed random variable with n experiments and p as the probability of success for each experiment is
    a. np
    b. np(1 – p)
    c. np/[np(1 – p)]
    d. n
    e. p

11. If we sample without replacement,
    a. it is important to consider the size of the sample relative to the size of the population.
    b. a larger sample relative to the size of the population is preferred because it will reduce the sampling error.
    c. the sample size is unimportant.
    d. use a smaller sample.
    e. we should take a census.

12. If we sample without replacement and the sample is large relative to the population,
    a. the sample mean will be small.
    b. the sample variance will be small.
    c. the sample mean will be large.
    d. the sample variance will be large.
    e. the sample variance must be adjusted.

13. If we sample with replacement,
    a. it is important to consider the size of the sample relative to the size of the population.
    b. a larger sample is preferred.
    c. the sample variance will not be biased.
    d. a smaller sample is preferred.
    e. the sample mean will be large.

14. If we sample with replacement and the sample is large relative to the population,
    a. the sample mean will be biased.
    b. the sample variance will be biased.
    c. the sample variance must be adjusted.
    d. the sample mean will be large.
    e. no adjustments need to be made.

True/False (If false, explain why)

1. The central limit theorem can be used to determine the distribution of the sample mean except distributions for population proportions.
2. To use the central limit theorem, we need to have a large enough sample size.
3. Results from a census will always be more accurate than results from a sample.
4. Only continuous distributions can use the central limit theorem.
5. The variance of the sample mean will decrease as the sample size increases.
6. The sample proportion can be found by dividing the number of sample members, x, by the sample size, n.
7. Sampling errors cannot be eliminated.
8. Simple random sampling is a procedure in which every element in the population has an equal chance of being selected.
9. In order to compute the mean and standard deviation of a sample, we must know the distribution of the sample.
10. The standard deviation of the sample mean becomes larger as the sample size increases.
11. Non-sampling errors can be eliminated by taking a census.
12. If we sample with replacement, we do not need to consider the size of the sample relative to the size of the population.
13. Sampling error is a type of systematic error.

Questions and Problems

1. Suppose you draw a sample of 100 from a population where the mean is 25,225 and the variance is 40,212. Find the mean of the sample mean and the variance of the sample mean.
2. Suppose 125 residents in Centerville, USA are asked whether they would approve the construction of a new high school. Of the 125 residents surveyed, 74 are in favor of the new school. Find the sample proportion and its variance.

3. Using the information in Problem 2, find the probability that the sample proportion is greater than .55.

4. From past history, a stockbroker knows that .05 of all people she "cold calls" will become new clients. Suppose she calls 200 people. What is the mean and variance of the sample proportion?

5. Using the information in Problem 4, find the probability that the sample proportion is less than .06.

6. Of the 200 economists who attended a conference, 40% are optimistic about the economy, whereas 60% are pessimistic about the economy. If 100 economists were sampled and asked whether they are optimistic about the economy, what kind of distribution is the sample proportion?

7. Of the 4,000 economists who attend a conference, 40% are optimistic about the economy and 60% are pessimistic about the economy. If 100 economists were sampled and asked whether they are optimistic about the economy, what kind of distribution is the sample proportion?

8. In Problem 7, what is the probability that between 50 and 60 economists asked are optimistic about the economy?

9. What is the probability of winning a game if the goal of the game is to score 10 or more on two rolls of a die?

Answers to Supplementary Exercises

Multiple Choice

1. b	6. d	11. a
2. a	7. b	12. e
3. e	8. b	13. c
4. b	9. a	14. e
5. a	10. b	

True/False

1. False. The sample proportion is a type of mean (average number of successes out of *n* experiments), and so the central limit theorem can also be used.
2. True
3. False. In some instances, a sample may be more accurate than a census.
4. False. The central limit theorem can also be applied to discrete distributions.
5. True
6. True
7. False. Sampling error can be eliminated by taking a census.
8. True
9. False. To compute the mean and standard deviation we do not need to know the distribution.
10. False. The standard deviation of the sample mean becomes smaller as the sample size increases.
11. False. Non-sampling error cannot be eliminated by taking a census. Only sampling error can be eliminated by taking a census.
12. True
13. False. Random error.

Questions and Problems

1. $E(\overline{x}) = 25{,}225$

   $Var(\overline{x}) = 4{,}021.2$

2. $\overline{p} = .592$

   $Var = .00193$

3. $P_r(> .55) = P_r(z > -0.9438) = 0.8264$

4. $E(\overline{p}) = .05$

   $\hat{\sigma}^2 = .0002375$

5. $P_r(\overline{p} < .06) = P_r[(\overline{p} - .05)/.0154 < (.06 - .05)/.154] = .5239$

6. The sampling proportion, $\bar{p} = x / n$, is distributed as

$$\bar{p} = \frac{x}{n} \sim N\left[0.4, \left(\sqrt{\frac{0.4(1-0.4)}{100}}\sqrt{\frac{200-100}{200-1}}\right)^2\right]$$

7. The sampling proportion, $\bar{p}$ = x/n, is distributed as

$$\bar{p} = \frac{x}{n} \sim N\left[0.4, \left(\sqrt{\frac{0.4(1-0.4)}{100}}\right)^2\right]$$

8. $$P_r\left(0.5 < \frac{x}{n} < 0.6\right) = P_r\left(\frac{0.5-0.4}{\sqrt{\frac{0.4(1-0.4)}{100}}} < z \quad \frac{0.6-0.4}{\frac{\sqrt{0.4(1-0.4)}}{100}}\right)$$

$$= P_r(2.04 < z < 4.08) = .0207$$

9. There are 36 possible outcomes in the game, but only six possible rolls that equal or exceed 10: (4, 6), (5, 5), (6, 4), (5, 6), (6, 5), and (6, 6). Therefore, there is a 6/36 chance of winning.

<u>**CHAPTER 9**</u>

OTHER CONTINUOUS DISTRIBUTIONS AND MOMENTS FOR DISTRIBUTIONS

Chapter Intuition

In Chapter 7, you learned about the normal distribution. Although the normal distribution can be used to approximate many types of data, there are many instances when the normal distribution will not be appropriate. In that case, we will need to use other distributions.

In this chapter, we will learn about two types of continuous distributions. The first type of distribution can be used to describe some population. The exponential and uniform distributions fall into this category. The second type of distribution is used to describe the distribution of a statistic and are known as sampling distributions. For example, if x is normally distributed with a mean of μ and an unknown standard deviation of σ, then the statistic (- μ)/(/n) follows a t distribution. So the t distribution describes the distribution of (- μ)/(/n), a statistic derived from the raw sample.

Chapter Review

1. The simplest of all continuous distributions is the ***uniform distribution***, where the random variable x is assumed to have equal probability of taking on any value over a given interval. For example, if x can only take on values between 5 and 10, and if it is equally likely that x will assume any value between 5 and 10 then x is uniformly distributed. Below is a graph of the uniform distribution.

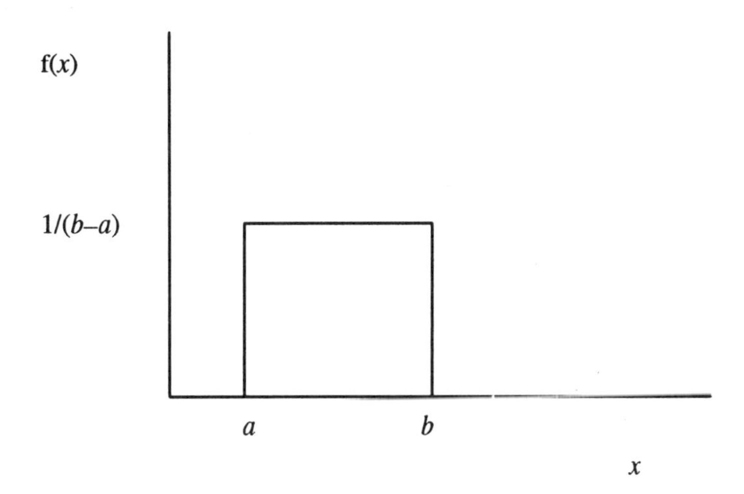

2. The ***exponential distribution*** is a continuous distribution which is closely related to the Poisson distribution. However, although the Poisson distribution is used to compute the probability of certain number of occurrences during a given interval, the exponential distribution is used to compute the likelihood of encountering specified intervals of time between consecutive occurrences.

3. ***Student's t distribution***, which is similar to the normal distribution, is used when the standard deviation of x is unknown; it is replaced by an estimator when the sample is small. Student's t distribution is described by its mean, variance, and ***degrees of freedom***. As the degrees of freedom get larger and larger, Student's t gets closer and closer to the normal distribution.

4. From the above figure, we can see that when the degree of freedom is small, the *t* distribution is shorter, with fatter tails than the normal distribution.

5. The ***chi-square distribution*** is a continuous distribution that is made up of the sum of squared independent standard normal random variables. The chi-square distribution which is positively skewed, is described by its degrees of freedom.

6. The ***F distribution*** is a continuous distribution which can be formed from the ratio of two independent chi-square distributions. The F distribution which is skewed, is described by two numbers, the degrees of freedom in the numerator and the degrees of freedom in the denominator.

7. ***Moments*** are measurements used to describe the properties of the distribution. Moments describe the central tendency, dispersion, symmetry, and peakedness of a distribution.

8. Continuous distributions like the uniform distribution and the exponential distribution can be used to describe a population. Sampling distributions can be used to test hypotheses about the population.

Useful Formulas

Uniform probability density function:

$$f(x) = \begin{cases} 1/(b-a) & \text{if } a \le x \le b \\ 0 & \text{otherwise} \end{cases}$$

Uniform cumulative distribution function:

$$F(x) = \begin{cases} 0 & \text{if } a \le x \le b \\ (x-b)/(b-a) & \text{if } x \le a \\ 1 & \text{if } x > b \end{cases}$$

Mean and variance for uniform distribution:

$$\mu_x = \frac{a+b}{2}, \sigma_x^2 = \frac{b-a}{\sqrt{12}}$$

Exponential probability density function:

$$f(x) = \begin{cases} \lambda e^{\lambda x} & \text{if } x \ge 0,\ \lambda > 0 \\ 0 & \text{if } x < 0 \end{cases}$$

Exponential cumulative distribution function:

$$F(x) = 1 - e^{-\lambda x}$$

Mean and variance for exponential distribution:

$$\mu_x = \frac{1}{\lambda}, \quad \sigma_x^2 = \frac{1}{\lambda^2}$$

Properties of exponential distribution:

$$P_r(x > a) = e^{-\lambda a}$$

$$P_r(x < b) = 1 - e^{-\lambda b}$$

$$P_r(a < x < b) = e^{-\lambda a} - e^{-\lambda b}$$

t statistic:

$$\frac{\bar{x} - \mu}{s_x / \sqrt{n}} \sim t$$

Chi-square distribution:

$$\frac{(n-1)S_x^2}{\sigma_x^2} = \frac{\Sigma(x_i - \bar{x})^2}{\sigma_x^2} \sim \chi_n^2$$

F distribution:

$$\frac{s_x^2/\sigma_x^2}{s_y^2/\sigma_y^2} \sim F$$

Example Problems

Example 1 **Uniform Distribution**

Suppose a random variable x can only take on values in the range from –3 to 6 and that the probability that the variable will assume any value within any interval in this range is the same as the probability that x will assume another value in another interval of similar width in the range. What is the distribution of x? Draw the probability density function for x.

Solution: Since all values of x within the interval have the same probability, x is distributed uniformly.

$$f(x)\begin{cases} 1/[6-(-3)] & \text{if } -3 \le x \le 6 \\ 0 & \text{otherwise} \end{cases}$$

Uniform Distribution

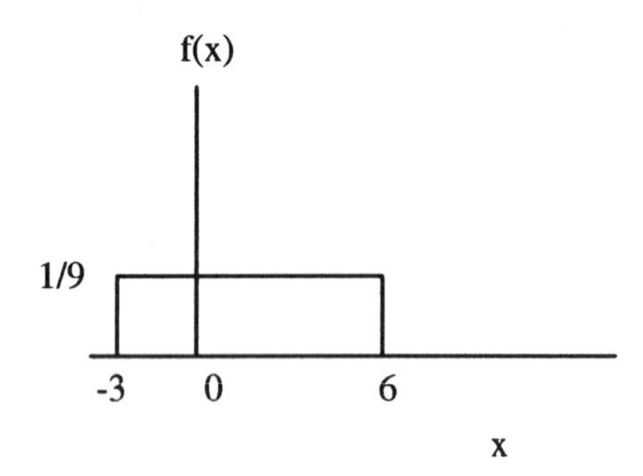

> Example 2 **Exponential Distribution**
>
> Suppose x has an exponential distribution $\mu_x = 3$. Find the following probabilities:
>
> a. $P_r(x < 2)$
>
> b. $P_r(x > 8)$
>
> c. $P_r(1 < x < 5)$

Solution: In order to use the exponential distribution we need to know the value of λ. Because the mean of the exponential distribution is $1/\lambda$, we can substitute $1/\mu_x$ for λ.

a. $P_r(x < 2) = 1 - e^{-(1/3)(2)}$
$= 1 - .5134$
$= .4866$

b. $P_r(x > 8) = e^{-(1/3)(8)}$
$= .0695$

c. $P_r(1 < x < 5) = e^{-(1/3)(1)} - e^{-(1/3)(5)}$
$= .7165 - .1889$
$= .5276$

Example 3 **t-Distribution**

Find the t_α for the following:

a. $\alpha = .05$ and df = 20

b. $\alpha = .025$ and df = 14

c. $\alpha = .10$ and df = 8

Solution: To solve this problem you need to look at the table in the text that provides the values for the t distribution.

a. $t_{.05,\,20} = 1.7247$

b. $t_{.025,\,14} = 2.1448$

c. $t_{.10,\,8} = 1.397$

Example 4 **Chi-square Distribution**

Find the following χ^2 values

a. $\alpha = .025$ and df = 5

b. $\alpha = .10$ and df = 45

c. $\alpha = .05$ and df = 10

Solution: To solve this problem, we use the values for the chi-square distribution listed in the text.

a. $\chi^2_{.025,\,5} = .831$

b. $\chi^2_{.10,\,45} = 33.350$

c. $\chi^2_{.05,\,10} = 3.940$

Example 5 **F-Distribution**

Find the following F values when:

a. $\alpha = .05$, $v_1 = 20$ and $v_2 = 5$

b. $\alpha = .01$, $v_1 = 5$ and $v_2 = 15$

c. $\alpha = .025$, $v_1 = 15$ and $v_2 = 5$

Solution: To solve this problem we use the table in the text where the values for the F distribution are given.

a. $F_{.05,\,20,\,10} = 2.77$

c. $F_{.01,\,5,\,15} = 4.56$

b. $F_{.01,\,15,\,5} = 9.72$

Example 6 **Uniform Distribution**

A professional association gives a standardized test to its new members every year. Suppose that the members' grades follow a uniform distribution with 98 points as the maximum and 55 points as the minimum.

a. Find the mean score.

b. Compute the standard deviation of the score.

c. If the passing grade is 75, what percentage of members will pass the course?

Solution:

a. mean of the uniform distribution = (a + b)/2

mean = (55 + 98)/2
= 76.5

b. variance of the uniform distribution = $(b - a)/\sqrt{12}$

variance = $(98 - 55)/\sqrt{12}$
= 12.41

standard deviation = $\sqrt{var}$
= $\sqrt{12.41}$
= 3.52

c. $P_r(x > c) = (b - c)/(b - a)$ for $a < c < b$
= (98 – 75)/(98 – 55)
= .5349 or 53.49%

Example 7 **Exponential Distribution**

The life of a certain model of car is found to be exponentially distributed with a mean of 95,000 miles. Find the proportion of cars which will have a life greater than 150,000 miles.

Solution: $P_r(x > 150{,}000) = e^{-(1/95{,}000)(150{,}000)}$
$= .2062$ or 20.62%

Supplementary Exercises

Multiple Choice

1. Sampling distributions
   a. describe a population.
   b. describe the distribution of a test statistic.
   c. must always have positive values.
   d. are always discrete.
   e. none of the above.
2. Suppose a random variable x follows a uniform distribution in the range from 2 to 7. The probability that the variable takes on a value of 9 is
   a. 1
   b. 0
   c. .5
   d. .25
   e. cannot be determined from the information given.
3. Suppose a random variable x follows a uniform distribution in the range from 2 to 7. The probability that the variable takes on a value of 9 or less is
   a. 1
   b. 0
   c. .5
   d. .25
   e. cannot be determined from the information given.

4. Suppose a random variable x is distributed normally with mean of μ and unknown standard deviation of σ. Then the distribution of (- μ)//n is
   a. t
   b. normal
   c. F
   d. χ^2
   e. binomial

5. The chi-square distribution
   a. is a continuous distribution described by its degrees of freedom.
   b. can be used to describe the distribution of the sample variance.
   c. is the sum of squared independent standard normal variables.
   d. is positively skewed.
   e. all of the above.

6. The F distribution is
   a. the ratio of two independent chi-square distributions divided by their respective degrees of freedom.
   b. the ratio of two binomial distributions.
   c. the ratio of two independent t distributions.
   d. the ratio of two independent normal distributions.
   e. always symmetric.

7. The exponential distribution is
   a. a discrete distribution.
   b. a continuous distribution determined by its parameter λ.
   c. the ratio of two independent chi-square distributions.
   d. the sum of squared independent normal variables.
   e. described by its degrees of freedom.

8. If x follows a uniform distribution in the range of 3 to 11, then the height of the probability density function is
   a. 11 + 3
   b. 11 – 3
   c. 1/(11 + 3)
   d. 1/(11 – 3)
   e. $(11 - 3)^2$

9. The mean for a uniform distribution in the range of a to b is
   a. (b – a)/2
   b. (a + b)/2
   c. (b – a)/√12
   d. (b + a)/√12
   e. $(b + a)^2$

10. The variance for a uniform distribution in the range of a to b is
   a. (b –a)/2
   b. (a + b)/2
   c. $(b - a)/\sqrt{12}$
   d. (b + a)/√12
   e. $(b + a)^2$

11. The probability density function for the exponential distribution, with parameter λ is
    a. $\lambda e^{-\lambda t}$
    b. λ
    c. $1 - e^{-\lambda t}$
    d. $1 + e^{-\lambda t}$
    e. $e^{-\lambda t}$

12. The cumulative distribution function for the exponential distribution, with parameter λ is
    a. $\lambda e^{-\lambda t}$
    b. λ
    c. $1 - e^{-\lambda t}$
    d. $1 + e^{-\lambda t}$
    e. $e^{-\lambda t}$

13. If x is a uniformly distributed random variable in the range of 1 to 9, then the mean of x is
    a. 1
    b. 9
    c. 5
    d. 2.31
    e. 4.5

14. If x is a uniformly distributed random variable in the range of 1 to 9, then the variance of x is
    a. 1
    b. 9
    c. 5
    d. 2.31
    e. 4.5

15. If x is a uniformly distributed random variable in the range of 1 to 9, then the probability that x takes on a value less than 5 is
    a. 1
    b. 1/2
    c. 1/4
    d. 3/4
    e. 0

16. If x is a uniformly distributed random variable in the range of 1 to 9, then the probability that x takes on a value greater than 3 is
    a. 1
    b. 1/2
    c. 1/4
    d. 3/4
    e. 0

17. If x is an exponentially distributed random variable with $\lambda = 2$, then the mean of x is
    a. 1
    b. 1/2
    c. 1/4
    d. 3/4
    e. 0

18. If x is an exponentially distributed random variable with $\lambda = 2$, then the variance of x is
    a. 1
    b. 1/2
    c. 1/4
    d. 3/4
    e. 0

19. If x is an exponentially distributed random variable with $\lambda = 2$, then the probability that x takes on a value less than 1/2 is
    a. .6321
    b. .3679
    c. .7358
    d. .2642
    e. 0

20. If x is an exponentially distributed random variable with $\lambda = 2$, then the probability that x takes on a value greater than 1/2 is
    a. .6321
    b. .3679
    c. .7358
    d. .2642
    e. 0

True/False (If false, explain why)

1. The t distribution is described by its mean, variance, and degrees of freedom.
2. The exponential distribution can be used to examine the probability of waiting time between consecutive occurrences.
3. If x is uniformly distributed in the range from 1 to 12, then the probability that x takes on a value less than 1 is 1.
4. The chi-square distribution is the sum of any squared independent random variables.
5. As the sample size increases, the t distribution gets closer to the normal distribution.
6. The mean and variance for the exponential distribution are both equal to λ.
7. The normal distribution is leptokurtic (coefficient of kurtosis > 3).
8. The chi-square distribution is a positively skewed distribution.
9. The F distribution is described by its mean and variance.
10. When x is uniformly distributed in the interval from a to b, x is more likely to to take on values above the mean than below the mean.

Questions and Problems

1. Suppose a random variable, x, follows a uniform distribution in the range of 25 to 37.

   a. Find the height of the probability density function.

   b. Find the probability that x will be less than 30.

   c. Find the probability that x will lie between 29 and 32.

2. Suppose x has an exponential distribution with $\mu_x = 1/2$. Find the following probabilities:

   a. $P_r(x < 1/2)$ b. $P_r(1/3 < x < 2/3)$

3. Find the t_α for the following:

   a. $\alpha = .10$ and df = 5 b. $\alpha = .01$ and df = 10

4. Find the χ^2 values when:

   a. $\alpha = .05$ and df = 30 b. $\alpha = .10$ and df = 20

5. Find the F values for the following:

   a. $\alpha = .05$, $\nu_1 = 10$ and $\nu_2 = 20$ c. $\alpha = .01$, $\nu_1 = 30$ and $\nu_2 = 20$

   b. $\alpha = .05$, $\nu_1 = 20$ and $\nu_2 = 10$

6. The Scholastic Aptitude Test (SAT) has a minimum score of 400 and a maximum score of 1600. Assume the scores follow a uniform distribution.

   a. Find the mean SAT score.

   b. Find the variance of the distribution.

   c. Find the percentage of students who will receive above 1200 on the test.

7. The life of Bright White Light Bulbs is found to be exponentially distributed with a mean of 2,500 hours. Find the proportion of light bulbs that will have a life greater than 3,000 hours.

8. A firm is interested in knowing the length of a warranty to issue. From previous experience, it knows that the products' life follows an exponential distribution with mean $1/\lambda = 10$ years. If the firm wants to be sure that no more than 10 percent of the products will be returned during the warranty period, what should be the length of the warranty?

Answers to Supplementary Exercises

Multiple Choice

1. b	6. a	11. a	16. d
2. b	7. b	12. c	17. b
3. a	8. d	13. c	18. b
4. a	9. b	14. d	19. a
5. e	10. c	15. b	20. b

True/False

1. True
2. True
3. False. The probability that x takes on a value outside its range is 0.
4. False. The chi-square distribution is the sum of squared independent standard normal random variables.
5. True
6. False. The mean and standard deviation for the exponential distribution are equal to $1/\lambda$.
7. False. The normal distribution is mesokurtic (CK = 3).
8. True
9. False. The F distribution is described by its degrees of freedom of its numerator and denominator.
10. False. If x is uniformly distributed, then all values in the interval are equally likely.

Questions and Problems

1. a. $1/(37 - 25)$

   b. $(30 - 25)/(37 - 25) = .417$

   c. $(32 - 25)/(37 - 25) - (29 - 25)/(37 - 25) = .25$

2. a. .632

   b. .250

3 a. 1.476

b. 2.764

4. a. 44.773

b. 28.412

5. a. 2.77

b. 2.35

c. 2.55

6. a. $(1600 + 400)/2 = 1000$

b. $(1600 - 400)/\sqrt{12} = 346.41$

c. $(1600 - 1200)/(1600 - 400) = .333 = 33.3\%$

7. .301

8. $P_r(x > a) = .90$, where a is the length of the warranty.

$P_r(x > a) = e^{-.1\,a} = .90$

Solving for a, we get a = 1.054 years.

CHAPTER 10

ESTIMATION AND STATISTICAL QUALITY CONTROL

Chapter Intuition

In Chapters 1-4, we learned about descriptive statistics, that is, statistics that can be used to describe a series of data. In Chapters 5-9, we learned how probability and probability distributions can be used to deal with uncertainty. In Chapter 10, we begin to learn about two subjects in inferential statistics. Chapter 10 concentrates on estimation, and Chapter 11 discusses hypothesis testing. Estimation deals with making "educated guesses" about unknown population parameters. For example, advertisers may be interested in knowing what percentage of T.V. viewers watch a certain television show. Because of the costs involved in surveying every television viewer, the actual percentage of viewers watching the show will never be known. However, by using the concepts of sampling, discussed in Chapter 8, and the concept of estimation discussed in this chapter, it will be possible to make an "educated guess" about the actual percentage of viewers.

Estimation is a "guess" of an unknown parameter such as the population mean or population proportion. Estimating a parameter is like shooting a gun at a target. The target is the parameter of interest, the estimator is your gun, and the data are your ammunition. Point estimation is like firing a single shot at the target; you only have one chance to hit the target. On the other hand, interval estimation is like firing a series of shots at a range of targets, using a machine gun. With interval estimation, we have a greater opportunity of hitting the target because we are firing at a range of targets.

In order to determine whether an estimator is good or not, we use the following criteria: unbiasedness, efficiency, and consistency. The unbiasedness criterion asks if the expected value of the test statistic is the true value. In our gun analogy, an unbiased estimator would be one where the scope on the rifle is aligned correctly. You may not always hit the target, but your misses should be scattered around the target, with your average shot being right at the target. On the other hand, a biased estimator would be like a gun whose scope is incorrectly aligned to the left of the target. In that case, most of your shots would finish to one side of the target and your shot pattern would be biased.

The efficiency criterion refers to the standard error (dispersion) of the statistic. If the estimation process is efficient, the standard error will be small. Using our gun and target analogy, an estimator is biased if the bullet holes are not "on average" around the target. On the other hand, efficiency deals with how scattered the bullet holes are from the target. If the bullet holes are far away from each other and far from the target, then the estimator is not efficient.

The final criterion we use is consistency. Consistency implies that the probability that the estimator will generate the true value gets closer to 1 as the sample size increases. This is an important property for a sample because it means that the time and money spent collecting more data is worthwhile because it improves the accuracy of our estimator.

Chapter Review

Because we are estimating the value of an uncertain population parameter, we can not be perfectly certain what the value of that parameter is. However, we can use the sampling and distribution theory discussed in previous chapters to provide a statistical interval for the true population parameter.

1. Calculation of the ***population parameters*** using samples is called ***estimation***. The formulas used to do the calculations are the ***estimators***. The numbers generated by the estimators are called the ***estimates***. A ***point estimate*** is a single number that is obtained from the estimator.
2. Four important criteria for evaluating estimators are:
   a. ***Unbiasedness***. An estimator is unbiased if its expected value equals the population

      $E(\hat{\theta}) = \theta$ -- that is if $E(\hat{\theta}) - \theta = 0$.

      parameter. Mathematically, an estimator is unbiased if
   b. ***Efficiency***. An estimator is efficient if it has a small variance.
   c. ***Consistency***. An estimator is consistent if the probability that the estimator will generate the true value approaches 1 as the sample size increases to infinity.
   d. ***Sufficiency***. A sufficient statistic is an estimator that utilizes all the information a sample contains about the parameters.
3. ***Interval estimation*** refers to the estimation of the parameter of interest with a ***confidence interval***. In other words, because the true value of the population parameter is unknown, we provide an interval that we believe may contain the true value. A 90 percent confidence interval implies that if the sampling procedure is repeated 100 times, we can expect the confidence interval to contain the true parameter 90 times. The width of the interval depends on several factors:
   a. the ***confidence level***. For example, if we want to be 99 percent certain that the interval contains the true value, we need a wider interval than if we only want to be 90 percent certain.

b. the variance of the population. If the original distribution has greater dispersion, the data will contain less information concerning the population parameter and the confidence interval will be wider.

c. the sample size. The greater the size of the sample, the more information we have and the smaller the width of the interval.

4. When the population variance is known, we use the standard normal distribution (*z*-distribution) to compute the confidence intervals for the mean.

5. When the population variance is unknown, we use Student's t distribution to compute the confidence interval for the mean.

6. Confidence intervals can also be computed for population proportions.

7. The chi-square distribution is used to compute the confidence interval for the population variance.

Useful Formulas

Unbiasedness:

$$E(\hat{\theta}) = \theta$$

Efficiency:

$$\text{Relative Efficiency} = \frac{V(\theta_2)}{V(\theta_1)}$$

Confidence intervals for the mean:

Variance known:

$$1-\alpha = P_r\left[\overline{X} - z_{\alpha/2}\left(\frac{\sigma}{\sqrt{n}}\right) < \mu < \overline{X} + z_{\alpha/2}\left(\frac{\sigma}{\sqrt{n}}\right)\right]$$

Variance unknown:

$$1-\alpha = P_r\left[\overline{X} - t_{n-1,\alpha/2}\left(\frac{s}{\sqrt{n}}\right) < \mu < \overline{X} + t_{n-1,\alpha/2}\left(\frac{s}{\sqrt{n}}\right)\right]$$

Confidence interval for a proportion:

$$1-\alpha = P_r\left[\hat{P}_x - z_{\alpha/2}\sqrt{\frac{\hat{P}_x(1-\hat{P}_x)}{n}} < P < \hat{P}_x + z_{\alpha/2}\sqrt{\frac{\hat{P}_x(1-\hat{P}_x)}{n}}\right]$$

Confidence interval for the variance:

$$1-\alpha - P_r\left[\frac{(n-1)s_x^2}{\chi^2_{n\text{-}1,\ \alpha/2}} < \sigma^2 < \frac{(n-1)s_x^2}{\chi^2_{n\text{-}1,\ (1\text{-}\alpha)/2}}\right]$$

Example Problems

Example 1 **Confidence Interval for the Population Mean**

A real estate agent in Wisconsin is interested in the mean home price in the state. A random sample of 40 homes shows a mean home price of $98,115 and a sample standard deviation of $27,100. Construct a 90% confidence interval for the mean home price.

Solution: When we construct a 90% confidence interval, we are interested in finding a range or interval in which the "true" mean home price would fall 90% of the time. The formula for constructing a confidence interval when the variance of the distribution is known is

$$\overline{x} = \pm z_{\alpha/2}\sigma / \sqrt{n}$$

For a 90% confidence interval, $\alpha = .10$, and so $z_{\alpha/2} = 1.645$.

$$98{,}115 \pm 1.645(27{,}100/0)$$

So the interval is from $91,066 to $105,164.

Example 2 **Confidence Interval for Known and Unknown Standard Deviation**

A quality control expert believes that the life of her firm's tires is normally distributed with a standard deviation of 15,225 miles. A random sample of 10 tires gives the following mileage on the life of the tires.

55,000 48,000 73,000 51,000 77,000 52,000 38,000 41,000 68,000 62,000

Construct a 95% confidence interval for the mean life of the tires.

Solution: First we need to find the mean life of our sample, .

$$\begin{aligned}\bar{x} &= (55{,}000 + 48{,}000 + 73{,}000 + 51{,}000 \\ &\quad + 77{,}000 + 52{,}000 + 38{,}000 + 41{,}000 \\ &\quad + 68{,}000 + 62{,}000) / 10 \\ &= 56{,}500\end{aligned}$$

When the sample is small (less than 30) and the standard deviation is unknown, we can no longer use the normal distribution to construct our confidence intervals. In this case, we need to use the *t* distribution. The formula for a confidence interval assuming a small sample and an unknown variance is

$$\bar{x} \pm t_{n-1,\alpha/2} s / \sqrt{n}$$

where

$$
\begin{aligned}
s^2 &= [(55{,}000 - 56{,}500)^2 + (48{,}000 - 56{,}500)^2 \\
&\quad + (73{,}000 - 56{,}500)^2 + (51{,}000 - 56{,}500)^2 \\
&\quad + (77{,}000 - 56{,}500)^2 + (52{,}000 - 56{,}500)^2 \\
&\quad + (38{,}000 - 56{,}500)^2 + (41{,}000 - 56{,}500)^2 \\
&\quad + (68{,}000 - 56{,}500)^2 + (62{,}000 - 56{,}500)^2]/9 \\
&= 173{,}611{,}111
\end{aligned}
$$

$$s = 13{,}176$$

$$56{,}500 \pm 2.26(13{,}176/0)$$

So the interval is from 47,083 to 65,916.

Example 3 **Confidence Interval for Sample Proportion**

A random sample of 500 residents of a city of 7,150 residents shows that 72% believe that the police commissioner is doing a good job of fighting crime. Construct a 99% confidence interval for the proportion of all residents that believe that the police commissioner is doing a good job against crime.

Solution: In this example, we want to find the confidence interval for a proportion. The formula for the confidence interval for a proportion is

$$\overline{p} \pm z_{\alpha/2}(\overline{p}(1-\overline{p})/n)^{1/2}$$

$$.72 \pm 2.81(.72(1 - .72)/500)^{1/2}$$

So the interval is from .6635 to .7764

Example 4 **Point Estimation**

Construct point estimates from the following situations:

a. A labor union randomly samples 50 of its members and finds that 29 *oppose* the new contract. Estimate the proportion of all workers who *favor* the new contract.

b. A statistics professor randomly samples 75 students in her class and finds that 42 do not know the meaning of a confidence interval. Estimate the proportion of all students in her class who cannot define a confidence interval.

Solution: a. Since 29 of the 50 workers oppose the contract, 21 must favor the contract. The proportion favoring the contract is 21/50 = .42.

b. A point estimate for the proportion of students who do not know the meaning of a confidence interval is 42/75 = .56.

Example 5 **Confidence Intervals**

A random sample of 50 observations from a population yielded the following summary statistics:

$$\Sigma x = 825 \qquad \Sigma x^2 = 22{,}242$$

Construct a 95% confidence interval for the population mean μ.

Solution: To solve this problem we need to compute the sample mean and sample standard deviation. Even though the population standard deviation is not known, we can use the normal distribution because we have a large enough sample.

$$\bar{x} = \Sigma x / n = 825 / 50 = 16.5$$

$s^2 = [\Sigma x^2 - (\Sigma x)^2/n]/(n-1)$
$= [22{,}242 - (825)^2/50]/(50-1)$
$= 176.11$

$s = 13.27$

$\bar{x} \pm z_{\alpha/2}\sigma/\sqrt{n}$

$16.5 \pm 1.96(13.27/0)$

So the interval is from 12.82 to 20.18.

Example 6 **Confidence Intervals**

A random sample of 200 observations from a population yielded the following summary statistics:

$\sum x_i = 1{,}500 \qquad \sum (x_i - \bar{x})^2 = 31{,}000$

Construct a 99% confidence interval for the population mean μ.

Solution:

$\bar{x} = 1{,}500/200 = 7.5$

$s^2 = 31{,}100/199 = 156.28$

$s = 12.5$

$\bar{x} \pm z_{\alpha/2}\sigma/\sqrt{n}$

$7.5 \pm 2.575(12.5/0)$

So the interval is from 5.22 to 9.78.

Example 7 **Confidence Interval for Population Variance**

Suppose a random sample of 51 boxes of cereal is taken and the sample variance is found to be 4.3 ounces. Construct a 90% confidence interval for the population variance.

The formula for the confidence interval for the variance is

$$1-\alpha = P_r\left[\frac{(n-1)s_x^2}{\chi^2_{n-1,\alpha/2}} < \sigma^2 < \frac{(n-1)s_x^2}{\chi^2_{n-1,(1-\alpha)/2}}\right]$$

Solution:

[(51 – 1)4.3]/71.42 to [(51 – 1)4.3]/32.57

So the interval is from 3.01 to 6.60.

Example 8 **Sample Size**

The manager of a local soft drink factory wants to construct a 95% confidence interval to estimate the average amount of soda pumped into the 12-ounce cans. From previous experience, he is confident that the standard deviation of the soda is 0.02. If he wants to control the width of the confidence interval to ± 0.005, how many cans of soda should he sample in each experiment?

Solution:

The 95% confidence interval is

$$\bar{x} \pm z_{\alpha/2}\sigma / \sqrt{n}$$

$$z_{\alpha/2}\sigma / \sqrt{n} = 1.96(.02 / \sqrt{n}) = .005$$

Solving for n yields n = 61.47. Therefore, he should collect 62 cans for each experiment.

Example 9 **Sample Size for a Proportion**

A local newspaper wants to estimate the proportion of voters who favor a tax increase to fund a recycling project. The editor wants to estimate the proportion with a 95% confidence interval. In addition, she wants the error margin to be within 3%. How large a sample should she take?

Solution: The error margin is

$$\pm z_{\alpha/2}\sqrt{\frac{p(1-p)}{n}}$$

which should be no more than $\pm 3\%$

$z_{.05/2} = 1.96$

The proportion, p, that will generate the widest interval (worst case scenario) is 1/2, so,

$$1.96\sqrt{\frac{1/2(1-1/2)}{n}} = 3\%$$

Solving for n yields n = 1076.1. Therefore, she should sample at least 1068 voters.

Supplementary Exercises

Multiple Choice

1. A point estimate is
   a. a single value that is used to estimate the population parameter.
   b. a range of values used to estimate the population parameter.
   c. always unbiased.
   d. always efficient.
   e. sufficient to establish unbiasedness.

2. An interval estimate is
   a. a single value that is used to estimate the population parameter.
   b. a range of values used to estimate the population parameter.
   c. always unbiased.
   d. always efficient.
   e. always a sufficient statistic.

3. The larger the population variance the
   a. narrower the width of the confidence interval, other things being equal.
   b. wider the width of the confidence interval, other things being equal.
   c. larger the mean.
   d. smaller the mean.
   e. larger the point estimate.

4. Other things being equal, the width of a 90% confidence interval will be
   a. wider than a 95% confidence interval.
   b. narrower than a 95% confidence interval.
   c. may be wider or narrower depending on the population variance.
   d. the same width as a 95% confidence interval.
   e. wider than a 99% confidence interval.

5. An unbiased estimator
   a. has the smallest variance among all estimators.
   b. is always the best estimator.
   c. has an expected value equal to the true parameter value.
   d. always generates the true value of the parameter.
   e. is always sufficient.

6. An efficient estimator
   a. has a small variance.
   b. gets closer to the true parameter value as the sample size increases.
   c. has an expected value equal to the true parameter value.
   d. always generates the true value of the parameter.
   e. is always sufficient.

7. A consistent estimator
   a. has the smallest variance among all estimators.
   b. gets closer to the true parameter value as the sample size increases.
   c. has an expected value equal to the true parameter value.
   d. is always sufficient.
   e. is always the best estimator.

8. When constructing interval estimates for the population mean we should use the
   a. z distribution when the population variance is unknown and the t distribution when the population variance is known.
   b. z distribution when the population variance is known and the t distribution when the population variance is unknown and is estimated with a small sample.
   c. z distribution whether the population variance is known or not, if the sample size is large enough.
   d. chi-squared distribution.
   e. F distribution.

9. To construct the confidence interval for the population variance, we should use the
   a. normal distribution.
   b. t distribution.
   c. F distribution.
   d. chi-square distribution.
   e. binomial distribution.

10. A sufficient statistic
    a. is consistent.
    b. is unbiased.
    c. uses all information a sample contains about the parameter to be estimated.
    d. is always efficient.
    e. approaches the true parameter value as the sample size decreases.

11. The confidence interval for the population mean when the population variance is known is
    a. $P_r(x \leq z)$
    b. $\bar{x} \pm z_{\alpha/2}\sigma/\sqrt{n}$
    c. $P_r(x \geq z)$
    d. 1
    e. 0

12. The width of a $(1 - \alpha)$ confidence interval for a population mean is
    a. $2\sigma z_{\alpha/2}$
    b. $\sigma z_{\alpha/2}$
    c. $2\sigma/\sqrt{n}\, z_{\alpha/2}$
    d. $\sigma/\sqrt{n}\, z_{\alpha/2}$
    e. $2\sigma^2$

13. Suppose a pollster surveys 300 people to see if they favor prayer in school. Of the 300 people surveyed, 127 favor prayer in school. An estimator of the proportion of people who favor prayer in school would be
    a. 127
    b. 300
    c. 127/300
    d. 127/300[1 – (127 – 300)]/300
    e. 300/127

14. An estimate of the variance of people who favor prayer in school in question 13 would be
    a. 127
    b. 300
    c. 127/300
    d. 127/300[1 – (127 – 300)]/300
    e. 300/127

15. Other things being equal, width of a confidence interval will be
    a. wider if the population variance is known.
    b. narrower if the population variance is unknown.
    c. wider if the population variance is unknown.
    d. the same width whether the population variance is known or not.
    e. wider if the mean is unknown.

True/False (If false, explain why)

1. Other things being equal, the greater the population variance, the narrower the width of the confidence interval.
2. For a given population variance and sample size, the greater the confidence level, the greater the width of the interval.
3. When the population variance is unknown and the sample size is small, we should use the t distribution to construct the confidence interval for the population mean.
4. When the population variance is unknown and the sample size is large, we should use the t distribution to construct the confidence interval.
5. Other things being equal, increasing the sample size increases the width of the confidence interval.
6. To construct a confidence interval for the population variance, we should use the chi-square distribution.
7. A confidence interval is really just a range of values constructed by sampling that will cover the true value $1 - \alpha$ percent of the time.
8. A larger population variance increases the width of the confidence interval because there is more dispersion around the population mean.
9. We can obtain an unbiased estimate of mean earnings per person in a city by randomly sampling people who live on the same street.
10. Consistent estimators reduce the need to increase the sample size.
11. To construct a $1 - \alpha$ confidence interval for the population mean with a known population variance, we should use z_α.
12. To construct a $1 - \alpha$ confidence interval for the population mean with an unknown population variance, we should use t_α.
13. When the population variance is unknown, it is inappropriate to use the standard normal distribution to construct a confidence interval.
14. A sufficient estimator utilizes all the information a sample contains about the parameters.

Questions and Problems

1. Suppose a golfer is interested in the average distance he hits a five iron. He hits 100 shots with the five iron and finds the sample mean to be 175 yards. If the population variance for his shots is known to be 64, construct a 95% confidence interval for the population mean.
2. Suppose a golfer is interested in the average distance he hits a five iron. He hits 25 shots with the five iron and finds the sample mean to be 175 yards. Assume that the population variance for his shots is unknown, but the sample variance is estimated to be 64, construct a 95% confidence interval for the population mean.
3. A random sample of 200 residents of Home Town, U.S.A. shows that 65% believe the superintendent of schools is doing a good job. Construct a 95% confidence interval for the proportion of all residents that believe the superintendent is doing a good job.
4. Suppose a college professor randomly samples 125 graduating seniors and finds that 47 have job offers. Estimate the proportion of students who have job offers.
5. A random sample of 30 radial tires is taken and found to have sample standard deviation of 2,352 miles. Construct a 95% confidence interval for the population variance.

Answers to Supplementary Exercises

Multiple Choice

1.	a	6.	a	11.	b
2.	b	7.	b	12.	c
3.	b	8.	d	13.	c
4.	b	9.	d	14.	d
5.	c	10.	c	15.	c

True/False

1. False. The confidence interval will be wider.
2. False. The confidence interval will be narrower.
3. True

4. False. When the sample size is large enough, we can also use the z-distribution, although the t-distribution is also correct.
5. False. The confidence interval gets narrower.
6. True
7. True
8. True
9. False. Because people with similar incomes tend to live in the same neighborhood, the mean earnings will be biased.
10. False. The advantage of a consistent estimator appears when the sample size is large.
11. False. We should use $z_{\alpha/2}$.
12. False. We should use $t_{\alpha/2}$.
13. False. If the sample size is large enough, we can use the standard normal distribution even if the population variance is unknown.
14. True

Questions and Problems

1.

$$\bar{x} \pm z_{\alpha/2} \frac{\sigma}{\sqrt{n}}$$

$$175 \pm 1.96 \frac{8}{\sqrt{100}}$$

So the interval is from 173.43 to 176.57.

2. Because the sample size is small and the variance is unknown, we must use the t-distribution to construct the confidence interval.

$$\bar{x} \pm t_{n-1,\alpha/2} \frac{s}{\sqrt{n}}$$

$$175 \pm 2.064 \frac{8}{\sqrt{25}}$$

So the interval is from 171.70 to 178.30.

3. $$\bar{p} \pm z_{\alpha/2}\left(\sqrt{\frac{\bar{p}(1-\bar{p})}{n}}\right)$$

$$.65 \pm 1.96\left(\sqrt{\frac{.65(1-.65)}{200}}\right)$$

So the interval is from .584 to .716.

4. $\hat{p} = 47/125 = .376$

5. $$1-\alpha = P_r\left[\frac{(n-1)s_x^2}{\chi^2_{n-1,\,\alpha/2}} < \sigma^2 < \frac{(n-1)s_x^2}{\chi^2_{n-1,\,(1-\alpha)/2}}\right]$$

$$\left(\frac{(30-1)(2352^2)}{42.557} \quad \frac{(30-1)(2352^2)}{17.708}\right)$$

So the interval is from 3,769,655 to 9,059,477.

CHAPTER 11

HYPOTHESIS TESTING

Chapter Intuition

The second part of inferential statistics is ***hypothesis testing***. In hypothesis testing, unlike estimation, we are not interested in making an educated guess about the true value of some parameter. Instead, we are interested in deciding if enough evidence exists to overthrow what we believe is true.

This concept can best be understood by the following example. Suppose a scientific study indicates that cigarettes containing more than 20 units of nicotine, are extremely dangerous to smokers. Our hypothesis is that the average amount of nicotine in cigarettes is 20 units or more. In estimation, we would collect a sample to "guess" the average nicotine content. In hypothesis testing, we are less interested in the average nicotine content. Instead, we are interested in collecting enough evidence to decide whether cigarettes contain 20 units of nicotine or not. Because a high nicotine content is dangerous, the null hypothesis is $\mu \geq 20$. That is, we want to assume that the cigarettes have a dangerous level of nicotine, unless we have sufficient evidence to determine that the null hypothesis is wrong. To conduct this test, we compute the mean level of nicotine for a sample of cigarettes. If the sample average is much smaller than 20, we have enough evidence to say that the null hypothesis $\mu \geq 20$ is wrong. Otherwise, we will accept the null hypothesis.

When conducting a hypothesis test, we consider two different hypotheses, the ***null hypothesis*** and the ***alternative hypothesis***. Which should be the null and which should be the alternative hypothesis is determined by the costs associated with accepting the wrong hypothesis. A hypothesis test should be set up so that the null hypothesis represents the "status quo" or the hypothesis with the least cost if it is wrongly accepted. From our previous example, the null hypothesis is that cigarettes have a dangerous level of nicotine unless we can find sufficient evidence to say the null hypothesis is wrong. In other words, we assume the null hypothesis is true unless we have enough evidence to reject it.

Chapter Review

1. ***Hypothesis testing*** is a systematic procedure used in testing the correctness of assumptions made about a population parameter. The ***null hypothesis*** is the hypothesis which conforms to the status quo and is what we are trying to disprove. In other words, if we do not have enough evidence to reject the null hypothesis, we are willing to live with it. For example, in our legal system, the null hypothesis is that a person is innocent. If we do not have enough evidence to show the person is guilty "beyond a reasonable doubt," we allow the person to go free. The ***alternative hypothesis*** is the hypothesis we are trying to prove true if we have substantial evidence to reject the null hypothesis.
2. In conducting hypothesis testing, it is possible to make two kinds of errors. If we reject the null hypothesis when it is actually true, we have made a ***type I error***. If we accept the null hypothesis when it is actually false, we have made a ***type II error***.
3. There are several ways to set up an alternative hypothesis. If we specify only one value for the population parameter, we are using a ***simple hypothesis***. If we set up a range of values for the population parameter, we are using a ***composite hypothesis***.
4. Steps for hypothesis testing:
   a. Set up the null and alternative hypotheses. The status quo statement should be the null hypothesis. The statement we are trying to prove should be the alternative hypothesis.
   b. Study the parameter of interest and select an appropriate statistic to conduct the test.
   c. Carefully examine the type I and type II errors, then decide on an α level (chance of making a type I error). Check the appropriate probability table to obtain the critical value. How large an α level we choose depends on the costs of making a type I error.
   d. Compute the test statistic chosen in step (b) and compare it with the critical value in step (c) in order to determine if we accept or reject the null hypothesis.
5. The ***power of a test*** refers to the ability of the test to reject the null hypothesis. The more powerful a test, the more likely it is to reject the null hypothesis.
6. ***One-tailed tests*** are used when the range of values for rejecting the null hypothesis lies entirely on one side of the null hypothesis values. For example, if we were to test the null hypothesis that the average nicotine content is greater than or equal to 20 units, the evidence that leads to the rejection of the null hypothesis should fall entirely on the extreme left-hand side of the null hypothesis value.

7. ***Two-tailed tests*** are used when the range of values for the alternative hypothesis can lie on either side of the null hypothesis values. For example, if we are interested in whether a coin is fair, then getting a percentage of heads which is either significantly higher or significantly lower than 50% will constitute enough evidence to reject the null hypothesis that p = 50%.

8. In deciding which test statistic is appropriate for the test you are performing, the following rules apply:

   a. To conduct a test on the mean we use the z-statistic or Student's t-statistic depending on the following:

      i. If the population variance is known or if the sample size is large ($n > 30$), we use the z-statistic.

      ii. If the population variance is unknown and the sample is small ($n < 30$), we use the t-distribution.

   b. To test that the variance is equal to some value, use the chi-squared statistic.

   c. To test the equality of two variances, use the F-statistic.

Useful Formulas

Chi-square distribution:

$$\frac{(n-1)s_x^2}{\sigma_x^2} \sim \chi^2_{n-1}$$

Testing one mean-known variance:

$$\frac{\bar{x}-\mu_0}{\frac{\sigma_x}{\sqrt{n}}} \sim z$$

Testing one mean-unknown variance with large sample:

$$\frac{\bar{x}-\mu_0}{\frac{s_x}{\sqrt{n}}} \sim z$$

Testing one mean-unknown variance with small sample:

$$\frac{\bar{x}-\mu_0}{\frac{s_x}{\sqrt{n}}} \sim t_{n-1}$$

Testing a proportion:

$$\frac{\bar{p}-p}{\sqrt{\frac{p(1-p)}{n}}} \sim z$$

Testing the difference between two means-small sample:

$$\frac{(\bar{x}_1 - \bar{x}_2) - (\mu_1 - \mu_2)}{\sqrt{\frac{(n_1 - 1)s_1^2 + (n_2 - 1)s_2^2}{(n_1 - n_2 - 2)}\left(\frac{1}{n_1} + \frac{1}{n_2}\right)}} = t$$

$$df = n_1 + n_2 - 2$$

Testing the difference between two means-large sample:

$$\frac{(\bar{x}_1 - \bar{x}_2) - (\mu_1 - \mu_2)}{\sqrt{\frac{\sigma_1^2}{n_1} + \frac{\sigma_2^2}{n_2}}} = z$$

Example Problems

Example 1 **Critical Values for Standard Normal Distribution**

Find the critical values for the following standard normal distributions.

a. Two-tailed test for $\alpha = .01$

b. One-tailed test for $\alpha = .025$

c. Two-tailed test for $\alpha = .05$

d. One-tailed test for $\alpha = .01$

e. Two-tailed test for $\alpha = .10$

Solution: For a two-tailed test we want the critical value for $z_{\alpha/2}$.

a. $z_{.01/2} = z_{.005} = 2.57$

b. $z_{.025} = 1.96$

c. $z_{.05/2} = z_{.025} = 1.96$

d. $z_{.01} = 2.33$

e. $z_{.10/2} = z_{.05} = 1.645$

Example 2 **One-tailed Test for the Mean**

A college professor randomly selects 25 freshmen to study the mathematical background of the incoming freshman class. The average SAT score of these 25 students is 565 and the standard deviation is estimated to be 40. Using this information, can the professor reject the null hypothesis that the average test score is 550 or less? Use a 5% level of significance.

Solution:

Step 1: Set up the hypotheses.

H_0: $\mu \leq 550$

H_1: $\mu > 550$

Step 2: Select a test statistic.

$$\frac{\overline{x} - \mu}{s / \sqrt{n}} = t \qquad df = n - 1 = 25 - 1 = 24$$

Step 3: Establish the rejection area.

From the *t*-table, we can determine that $t_{5\%, 24} = 1.71$.

Step 4: Carry out the test

$$\frac{565-550}{40/\sqrt{25}} = 1.875 > 1.71 \quad \text{Reject } H_0$$

Note: In step 2, we used a t-test because the standard deviation was *unknown* and was estimated with a *small* sample (n = 25). If $n - 1 \geq 30$, we may use the normal distribution to approximate the t-distribution.

Example 3 **Testing the Difference of Two Means-Small Sample Case**

Suppose the college professor in Example 2 wants to know whether the mathematical aptitude of freshmen has improved over the past year. He randomly draws 25 SAT mathematics scores from last year's freshman class and obtains a sample mean of 560 and a sample standard deviation of 35. Compare the results from last year with the results from this year. Do the data offer enough evidence for rejecting the null hypothesis that there is no improvement? Use a 5% level of significance.

Solution:

Step 1: Set up the hypotheses

H_0: $\mu_{new} - \mu_{old} \leq 0$

H_1: $\mu_{new} - \mu_{old} > 0$

Step 2: Select a test statistic.

$$\frac{(\bar{x}_{new} - \bar{x}_{old}) - (\mu_{new} - \mu_{old})}{\sqrt{\dfrac{(n_{new}-1)s_{new}^2 + (n_{old}-1)s_{old}^2}{(n_{new}+n_{old}-2)}\left(\dfrac{1}{n_{new}} + \dfrac{1}{n_{old}}\right)}} = t$$

$$df = 25 + 25 - 2 = 48$$

Step 3: Establish the rejection area.

From the *t*-table, we can determine that $t_{5\%, 48} = 1.64$.

Step 4: Carry out the test.

$$\frac{(565 - 560) - 0}{\sqrt{\frac{(25-1)40^2 + (25-1)35^2}{(25+25-2)}\left(\frac{1}{25} + \frac{1}{25}\right)}} = 0.47 < 1.64$$

The evidence is not strong enough to reject the null hypothesis.
Note: The *t*-test is used because $n_{new} = n_{old} = 25$ and these are small samples.

Example 4 **Test of the Difference Between Two Means-Large Sample**

In example 3, assume that the college professor obtained a grant to study whether there was an improvement in the mathematics ability of freshmen. Using this grant money, he is able to obtain larger samples. The results are compiled in the following table.

	last year	this year
size of sample	125	100
average score	562	568
standard deviation	40	42

Can the professor reject the null hypothesis that the average scores improved? Use a 5% level of significance.

Solution:

Step 1: Set up the hypotheses.

H_0: $\mu_{new} - \mu_{old} \le 0$

H_1: $\mu_{new} - \mu_{old} > 0$

Step 2: Select a test statistic.

$$\frac{(\overline{x}_{new} - \overline{x}_{old}) - (\mu_{new} - \mu_{old})}{\sqrt{\frac{\sigma^2_{new}}{n_{new}} + \frac{\sigma^2_{old}}{n_{old}}}} = z$$

Step 3: Establish the rejection area.

From the standard normal table, we can determine that $z_{5\%} = 1.64$.

Step 4: Carry out the test.

$$\frac{(568 - 562) - 0}{\sqrt{\frac{40^2}{125} + \frac{42^2}{100}}} = 1.087 < 1.64$$

The evidence is not strong enough to reject the null hypothesis.

Example 5 **Test of a Proportion**

In a recent poll, 800 voters were randomly sampled and asked whether they would approve of the building of an incinerator in the state. According to the poll, 450 voters approved and 350 voters disapproved. Is there enough evidence to argue that more than 50% of the voters support the building of the incinerator? Use a 5% level of significance.

Solution:

Step 1: Set up the hypotheses.

H_0: $p \leq 0.5$

H_1: $p > 0.5$

Step 2: Select a test statistic.

$$\frac{\bar{p}-p}{\sqrt{\frac{p(1-p)}{n}}} \sim z$$

Step 3: Establish the rejection region.

From the standard normal table, we can determine that $z_{5\%} = 1.64$.

Step 4:

$$\frac{450/800-0.5}{\sqrt{\frac{0.5(1-0.5)}{800}}} = 3.33 > 1.64$$

So we can reject the null hypothesis.

Example 6 **Testing the Equality of Two Proportions**

A survey shows that 360 out of 500 females and 600 out of 800 males enjoy a game show. Based on this evidence, would you argue that the ratio of female and male viewers who enjoy the game show differs? Use a 5% level of significance.

Solution:

Step 1: Set up the hypotheses.

H_0: $p_f - p_m = 0$

H_1: $p_f - p_m \neq 0$

Step 2: Select a test statistic.

$$\frac{(\bar{p}_f - \bar{p}_m) - (p_f - p_m)}{\sqrt{p_c(1-p_c)\left(\frac{1}{n_f} + \frac{1}{n_m}\right)}} = z, \quad \text{where } p_c = \text{ combined proportion}$$

Step 3: Establish the rejection region.

This is a two-tailed test so we use $z_{\alpha/2}$ as our critical value. From the normal distribution table $z_{.025} = 1.96$. So we will reject the null hypothesis if our statistic $|z| > 1.96$.

Step 4: Carry out the test.

$$p_c = \frac{360 + 600}{500 + 800} = 0.74$$

$$\frac{(360/500 - 600/800) - 0}{\sqrt{0.74(1-0.74)\left(\frac{1}{500} + \frac{1}{800}\right)}} = -1.2 > -1.96$$

There is not enough evidence to reject the null hypothesis.

Example 7 **Test of the Variance**

The Yum-Yum Company owns a machine that dispenses dog food into 16 ounce cans. The quality of the machine is judged by the variance of the dog food dispensed into the cans. The quality control manager of the company collects 30 cans of dog food and weighs their contents. She estimates the variance to be 1 ounce. From this evidence, will she be able to claim that the machine has a variance higher than 0.5? Use a 5% level of significance.

Solution:

Step 1: Set up the hypotheses.

H_0: $\sigma^2 \leq 0.5$

H_1: $\sigma^2 > 0.5$

Step 2: Select a test statistic.

$$\frac{(n-1)s_x^2}{\sigma_x^2} = \chi^2_{n-1}$$

Step 3: Establish the rejection region.

From the χ^2 table the critical value for 5% and 29 degrees of freedom is 42.557.

Step 4: Carry out the test.

$$\frac{(30-1)1^2}{0.5} = 58 > 42.557$$

So we can reject the null hypothesis.

Example 8 **Testing the Equality of Two Variances**

Suppose the quality control manager in Example 7 is not satisfied with the packaging machine. She is thinking of replacing it with a better one. She tests a new machine that offers a 30-day money-back guarantee. A sample of 30 cans from the new machine has a variance of 0.8. Is there enough evidence to argue that the new machine has a smaller variance than the old machine?

Solution:

Step 1: Set up the hypotheses.

H_0: $\sigma_{old} \leq \sigma_{new}$

H_1: $\sigma_{old} > \sigma_{new}$

Step 2: Select a test statistic.

$$\frac{s^2_{old}}{s^2_{new}} = F_{29,29}$$

Step 3: Establish the rejection region.

From the F-distribution table, the critical value for $\alpha = 5\%$ and 29 degrees of freedom in the numerator and 29 degrees of freedom in the denominator is approximately 1.85.

Step 4: Carry out the test.

1/0.8 = 1.25

There is not enough evidence to reject the null hypothesis.

Example 9 **The Power of a Test for a Means Test**

You are working for a light bulb manufacturer. Your job is to make sure that the average life of the bulbs produced is greater than 300 hours. You randomly select 60 bulbs from the bulbs produced last week and find the sample average is 325 and the sample standard deviation is 32. Using this information, you test the null hypothesis that the mean life is less than 300 hours at the 5% level of significance. What is the power of the test when the true mean life is 312?

Solution:

Step 1: Obtain the rejection criteria.

Assume that the critical value for rejecting the null hypothesis of $\mu = 300$ is $\overline{x}_c$
Then

$$\frac{\overline{x}_c - 300}{32/\sqrt{60}} = 1.64$$

We can solve for $\overline{x}_c = 300 + 32\sqrt{60}\,(1.64) = 306.78$.

Step 2: Obtain the probability of rejecting H_0 assuming $\mu = 312$.

$$P_r(\overline{x} > 306.78 \mid \mu = 312)$$
$$= P_r\left(\frac{\overline{x} - 312}{32/\sqrt{60}} > \frac{306.78 - 312}{32/\sqrt{60}}\right)$$
$$= P_r(z > -1.26) = .8962$$

Example 10 **The Power of a Test for a Proportion**

You are working for a furniture manufacturer. Your job is to make sure that the springs purchased are not stronger than 40 pounds (to keep the springs from poking through the fabric). Each time a truckload of springs arrives, you check 600 springs. You will then test the null hypothesis that at least 2% of the springs are bad using a 5% level of significance. One day a truckload arrives that actually contains only 1.5% bad springs. What is the probability of taking this shipment? [Note: This is a power of a test question because we are interested in the probability of rejecting the null hypothesis.]

Solution:

Step 1: Find the critical value for $\overline{p}_c$

$$\frac{\overline{p}_c - 0.02}{\sqrt{\frac{0.02(1-0.02)}{600}}} = 1.64, \text{ solving for } \overline{p}_c = 0.0106$$

Step 2: Obtain the probability of rejecting the null hypothesis.

$$P_r(\overline{p} < 0.0106 \mid p = 0.015)$$

$$= P_r\left(\frac{\overline{p} - 0.015}{\sqrt{\frac{0.015(1-0.015)}{600}}} < \frac{0.0106 - 0.015}{\sqrt{\frac{0.015(1-0.015)}{600}}}\right)$$

$$= P_r(z < -0.88) = 0.19$$

Supplementary Exercises

Multiple Choice

1. A type I error
   a. results from accepting the null hypothesis when it is actually false.
   b. results from rejecting the null hypothesis when it is true.
   c. is a type of sampling error.
   d. represents the size of the test.
   e. is the difference between taking a sample and a census.

2. A type II error
   a. results from accepting the null hypothesis when it is actually false.
   b. results from rejecting the null hypothesis when it is true.
   c. is a type of sampling error.
   d. represents the size of the test.
   e. is the difference between taking a sample and a census.

3. The null hypothesis
   a. is the hypothesis we are trying to reject.
   b. conforms to the status quo.
   c. usually has the greatest cost when it is incorrectly rejected.
   d. results from a type I error.
   e. results from a type II error.

4. A one-tailed test of the population would be appropriate if
   a. H_1: $\mu \neq 0$
   b. H_0: $\mu = 0$
   c. H_1: $\mu > 0$
   d. all of the above hypotheses were true.
   e. would not be appropriate for any of the above hypotheses.

5. A two-tailed test of the population would be appropriate if
   a. H_1: $\mu \neq 0$
   b. H_1: $\mu < 0$
   c. H_0: $\mu > 0$
   d. all of the above hypotheses were true.
   e. would not be appropriate for any of the above hypotheses.

6. A simple hypothesis is
   a. another name for the null hypothesis.
   b. another name for the alternative hypothesis.
   c. one where we specify only one value for the population parameter.
   d. a sampling error.
   e. one where we specify a range of values for the population parameter.

7. A composite hypothesis is
   a. another name for the null hypothesis.
   b. another name for the alternative hypothesis.
   c. one where we specify only one value for the population parameter, Θ.
   d. a sampling error.
   e. one where we specify a range of values for the population parameter, Θ.

8. The power of a test refers to the test's
   a. ability to control the type I error.
   b. significance level.
   c. ability to reject the null hypothesis.
   d. ability to control the type II error.
   e. ability to reject the alternative hypothesis.

9. For a one-tailed test using the normal distribution, a significance level of .10 would have a critical value of
   a. 1.645
   b. 1.96
   c. 1.282
   d. 2.575
   e. 1.382

10. For a two-tailed test using the normal distribution, a significance level of .10 would have a critical value of
    a. 1.645
    b. 1.96
    c. 1.282
    d. 2.575
    e. 1.382

11. For a one-tailed test using the normal distribution, a significance level of .05 would have a critical value of
    a. 1.645
    b. 1.96
    c. 1.282
    d. 2.575
    e. 1.382

12. For a two-tailed test using the normal distribution, a significance level of .05 would have a critical value of

    a. 1.645
    b. 1.96
    c. 1.282
    d. 2.575
    e. 1.382

13. For a one-tailed test using the *t*-distribution, a significance level of .05, and 20 degrees of freedom would have a critical value of

    a. 1.725
    b. 1.96
    c. 2.086
    d. 2.528
    e. 1.625

14. For a two-tailed test using the *t*-distribution, a significance level of .05, and 20 degrees of freedom would have a critical value of

    a. 1.725
    b. 1.96
    c. 1.625
    d. 2.528
    e. 2.086

True/False (if false, explain why)

1. The power of a test is $1 - \alpha$.
2. Other things being equal, a two-tailed test will always have a larger absolute critical value than a one-tailed test.
3. There are greater costs associated with making a type II error than with making a type I error.
4. A type I error is the P_r(reject H_0 | H_0 is true).
5. A type II error is the P_r(accept H_0 | H_0 is false).
6. A test with low power will lead to over rejection of the null hypothesis.
7. A two-tailed test is used when we want to prove that the population mean, μ, is not equal to a specified value of μ_0.
8. A one-tailed test is used when we want to prove that the population mean, μ, is either much larger or much smaller than a specified value of μ_0.
9. A simple hypothesis is one where we specify a range of values for the population parameter.
10. Higher p-values lead to rejection of H_0, whereas lower p-values lead to acceptance of H_0.

11. When conducting a hypothesis test with 25 observations and an *unknown variance*, we should use the z distribution.

12. Other things being equal, the critical value will always be smaller when the variance is unknown.

13. There is a tradeoff between the size and power of a test.

14. Testing for the H_0: $\sigma^2 \geq 10$ vs. H_1: $\sigma^2 < 10$, we should use the normal distribution.

15. When comparing the variances of two normal populations, we should use the normal distribution.

Questions and Problems

1. Find the critical value for a sample size of 10 and a known variance of 2, for a two-tailed test, given the following levels of significance.

   a. $\alpha = .10$
   b. $\alpha = .05$
   c. $\alpha = .01$

2. Find the critical value for a sample size of 10 and a known variance of 2, for a one-tailed test, given the following levels of significance.

   a. $\alpha = .10$
   b. $\alpha = .05$
   c. $\alpha = .01$

3. Find the critical value for a sample size of 10 and an *unknown* variance of 2, for a two-tailed test, given the following levels of significance.

   a. $\alpha = .10$
   b. $\alpha = .05$
   c. $\alpha = .01$

4. Find the critical value for a sample size of 10 and a *unknown* variance of 2, for a one-tailed test, given the following levels of significance.

   a. $\alpha = .10$
   b. $\alpha = .05$
   c. $\alpha = .01$

5. Suppose you are given the following information

   $\bar{x} = 100$ $\sigma^2 = 64$ $n = 30$

   Conduct the following hypothesis test at the .05 level of significance:

   H_0: $\mu = 75$ vs. H_1: $\mu > 75$

6. A sample of 25 students taking the GMATs at Business School U. has a sample mean of 525 and a sample standard deviation of 90. Test the hypothesis that the mean GMAT score of BSU students is equal to 500, against the alternative hypothesis that the mean GMAT score is higher than 500, at the .05 level of significance.

7. The Play Like a Pro Golf School claims that the variance of a student's golf score will be less than 6 strokes per round. A random sample of 30 students who took the course found the variance to be 4 strokes per round. Assuming a normal distribution, test the golf school's claim at the 5% level of significance.

8. Suppose we are interested in the proportion of people in a small town who favor prayer in school. A random sample of 100 people finds 45 in favor of prayer in school. Test the null hypothesis that p equals .50 against the alternative hypothesis that p is less than .50 at the .10 level of significance.

9. A sample of 100 students in a high school have a sample mean score of 505 on the English portion of the SATs. If the sample standard deviation is 110, test the hypothesis that the high school's mean SAT score is 480 against the alternative hypothesis that the school's mean SAT score is greater than 480, at the .10 level of significance.

10. A GMAT review course claims that the variance of test scores of its graduates is less than 100. A random sample of 30 students who took the course is taken and found to have a variance of 95. Assuming a normal distribution, test the review course's claim at the 5% level of significance.

Answers to Supplementary Exercises

Multiple Choice

1.	b	6.	c	11.	a
2.	a	7.	e	12.	b
3.	d	8.	c	13.	a
4.	c	9.	c	14.	e
5.	a	10.	a		

True/False

1. False. The power of the test is $1 - \beta$.
2. True
3. False. Because of the way we construct the null hypothesis, a type I error usually has greater costs than a type II error.
4. True
5. True
6. False. Low power will lead to an under rejection of the null hypothesis.
7. True
8. True
9. False. A simple hypothesis is one where we specify only one value for the parameter.
10. False. Lower p-values lead to rejection of H_0.
11. False. When the variance is unknown and the sample size is small, we should use the t-distribution.
12. False. Other things being equal, the critical value will be larger when the variance is unknown.
13. True
14. False. Use the chi-square distribution.
15. False. Use the F-distribution.

Questions and Problems

1. a. 1.645 c. 2.575

   b. 1.96

2. a. 1.282 c. 2.327

   b. 1.645

3. a. 1.833 c. 3.250

   b. 2.262

4. a. 1.383 c. 2.821

   b. 1.833

5. We compute z = 17.17 and the critical value for a one-tailed test with $\alpha = .05$ is 1.645, so we reject the null hypothesis.

6. We compute t = 1.389 and the critical value for a one-tailed test with $\alpha = .05$ and 24 degrees of freedom is 1.711 so we reject the null hypothesis.

7. We compute $\chi^2 = 19.33$, critical value with $\alpha = .05$ and 29 degrees of freedom is 42.557, so we cannot reject the null hypothesis.

8. We compute $z = -1.005$, critical value for a one-tailed test with $\alpha = .10$ is -1.28, so we cannot reject the null hypothesis.

9. $z = (505 - 480)/(110/\sqrt{100}) = 2.27$

   The critical value for $\alpha = .10$ is 1.28. Therefore, we can reject the null hypothesis that the mean SAT score is equal to 480.

10. $H_0: \sigma^2 \geq 100$ vs. $H_1: \sigma^2 < 100$

$$\begin{aligned} \chi^2_{n-1} &= [(n-1)s^2]/\sigma^2 \\ &= [(30-1)95]/100 \\ &= 27.55 \end{aligned}$$

   The critical value for $\alpha = .05$ and 29 degrees of freedom is 42.557. Therefore, we are unable to reject the null hypothesis.

CHAPTER 12

ANALYSIS OF VARIANCE AND CHI-SQUARE TESTS

Chapter Intuition

In Chapter 11, we learned how to test for differences in the means of two groups. When we are interested in comparing the means of more than two groups, we use a procedure known as analysis of variance (ANOVA). For example, suppose a factory operates three shifts. If the manager of the factory wants to test the null hypothesis of equal productivity for all three shifts, ANOVA represents the appropriate technique.

ANOVA uses an *F*-statistic to compare the means of three or more groups. To understand how ANOVA works, consider the numerator of the F-statistic, which measures the difference between the average productivity of each shift and the average productivity of all three shifts. If there is similar productivity in all three shifts, then the average productivity of each shift should be close to the overall average productivity and the numerator will be small. When there is different productivity in all three shifts, the numerator should be large.

Two-way ANOVA is a method for comparing three or more means while controlling for a second factor. For example, in the previous factory example, we are interested in the productivity of the three shifts. However, we also recognize that the number of years of experience of the employees may have a bearing on the productivity. Therefore, we must account for this factor in order to draw correct conclusions.

The second topic covered in this chapter is goodness of fit tests. Goodness of fit tests are used to test whether the data are generated from a presumed distribution. To conduct this test, we need to determine what the frequency distribution should be if the data are really generated from the presumed distribution. In order to determine if the data do come from the presumed distribution, we compare the expected frequency of the presumed distribution with the observed frequency from the data. If the observed frequencies are similar to the expected frequencies, we can conclude that the data do come from the presumed distribution. Otherwise, we must reject the null hypothesis and conclude that the data may come from some other distribution.

Chapter Review

1. When we are interested in knowing whether the means of three or more populations equal each other or not, we use ***analysis of variance*** (ANOVA). Testing for the equality of means can be useful for examining a wide variety of problems in business. For example, suppose a factory operates three shifts. If the manager wants to test the null hypothesis of equal productivity in the three shifts, we can use ANOVA to answer this question.

2. ANOVA tests the equality of more than two means by comparing the variances between groups with the variances within groups. The variation between groups is measured by the ***between treatments sum of squares*** (SSB). The SSB measures the variation among the population means of the treatment groups. The variation within each group is measured by the ***sum of squares within treatments*** (SSW). The SSW measures the variation within each treatment group. The total variation for the data is measured by the ***total sum of squares*** (TSS).

3. ***One-way ANOVA*** tests the equality of more than two means, assuming that only one factor determines the differences in the means. From our factory example, the manager assumes that only one factor, the shift, determines productivity.

4. ***Two-way ANOVA*** tests the equality of more than two means, assuming that more than one factor determines the differences in the means. For example, when the manager examines the differences in productivity of the three shifts, he may also wish to control for the experience of the workers in each shift.

5. To illustrate how one-way ANOVA works, let us return to our factory work example. If we assume that the means of the three shifts are the same, then most of the variation of the data results from the within-group variation (SSW) and there will be very little between-group variation (SSB), because all groups have the same mean. On the other hand, if the means of the three shifts are different, there will be a large between-group variation and a smaller within-group variation. By taking the ratio of SSW to SSB (with an adjustment for degrees of freedom), we can form an F-statistic that allows us to determine if the means are equal.

6. **Goodness of fit tests** are used to test whether the data are generated from an hypothesized distribution. The test compares the observed frequency (f_o) with the expected frequency (f_e). The expected frequency is obtained by assuming that the null hypothesis is true; that is, it is the frequency we expect to see if the null hypothesis is true. The observed frequency represents the frequency we actually observe. When the observed frequency (our evidence) is far enough away from the null hypothesis, the test statistic will generate a large value and lead to a rejection of the null hypothesis.

7. Another use of a goodness of fit test is to test if two variables are independent. To test the independence of two variables, we construct a ***contingency table*** and compare the expected frequency with the observed frequency.

Useful Formulas

Testing the equality of three or more means:

$$F = \frac{SSB/(m-1)}{SSW/(n-m)}$$

$$SSB = \sum_{j}^{m} n_j \; (\bar{x}_j - \bar{x})^2$$

$$SSW = \sum_{j=1}^{m} \sum_{i=1}^{nj} (x_{ij} - \bar{x}_j)^2$$

Between treatments sum of squares:

$$SSB = \sum_{j}^{m} n_j (\bar{x}_j - \bar{x})^2$$

Between factors sum of squares:

$$SSF = \sum_{i=1}^{I} JK(\bar{x}_i .. - \bar{x})^2$$

Comparing the difference between two population means:

$$t = \frac{(\bar{x}_1 - \bar{x}_2) - (\mu_1 - \mu_2)}{\sqrt{\frac{n_1 s_1^2 + n_2 s_2^2}{(n_1 + n_2 - 2)} \left(\frac{1}{n_1} + \frac{1}{n_2} \right)}}$$

Goodness of fit test:

$$\Sigma \; \frac{(f_o - f_e)^2}{f_e} \sim \chi^2$$

Two-way ANOVA:

Treatment effect:

$$\frac{SSB/(J-1)}{SSE/(I-1)(J-1)} \sim F_{(J-1),(I-1)(J-1)}$$

Factor effect:

$$\frac{SSF/(J-1)}{SSE/(I-1)(J-1)} \sim F_{(J-1),(I-1)(J-1)}$$

Example Problems

Example 1 **One-way ANOVA**

The following table shows the sales figures (in thousands) for salespersons with different years of experience. Can you reject the null hypothesis that the performances of the salespeople equal each other?

	Salesperson's Experience (years)		
	0 - 2	3 - 4	5 or more
Sales (000)	13	18	16
	17	17	17
	16	15	20
	18	14	19
	16	21	18
Average	16	17	18

Solution:

$\overline{x} = 17,\ \ \overline{x}_1 = 16,\ \overline{x}_2 = 17,\ \overline{x}_3 = 18$

	$(x_{i1} - \overline{x}_1)^2$	$(x_{i2} - \overline{x}_2)^2$	$(x_{i3} - \overline{x}_3)^2$
	$(13-16)^2$	$(18-17)^2$	$(16-18)^2$
	$(17-16)^2$	$(17-17)^2$	$(17-18)^2$
	$(16-16)^2$	$(15-17)^2$	$(20-18)^2$
	$(18-16)^2$	$(14-17)^2$	$(19-18)^2$
	$(16-16)^2$	$(21-17)^2$	$(18-18)^2$
Total	14	30	10

$$SSB = (16-17)^2 + (17-17)^2 + (18-17)^2 = 2$$

$$SSW = \sum(x_{i1} - \bar{x}_1)^2 + \sum(x_{i2} - \bar{x}_2)^2 + \sum(x_{i3} - \bar{x}_3)^2$$

$$\begin{aligned} F &= [SSB/(m-1)]/[SSW/(n-m)] \\ &= [2/(3-1)]/[54/(15-3)] \\ &= 0.2222 < F_{2,12} = 2.81 \end{aligned}$$

Example 2 **Two-way ANOVA**

We use the same data given in Example 1, except that we assume that the salespersons scored differently on an aptitude test given to them during their recruitment. The sales performance and aptitude scores are presented in the following table.

		Salesperson's Experience (years)			Row Means
		0 - 2	3 - 4	5 or more	
Scores	A	13	18	16	15.7
	B	17	17	17	17
	C	16	15	20	17
	D	18	14	19	17
	E	16	21	18	18.3
Column Means		16	17	17	17

Can you argue that the scores make a difference?

Solution:

$$SSB = \sum(\bar{x}_k - \bar{x})^2 = (16-17)^2 + (17-17)^2 + (18-17)^2 = 2$$

$$\begin{aligned} SSF &= \sum(\bar{x}_j - \bar{x})^2 \\ &= (15.7-17)^2 + (17-17)^2 + (17-17)^2 + (17-17)^2 + (18.3-17)^2 \\ &= 3.56 \end{aligned}$$

$SST = (13 - 17)^2 + (17 - 17)^2 + (16 - 17)^2 + \ldots + (18 - 17)^2 = 64$

$SSE = SST - SSB - SSF = 64 - 2 - 3.56 = 58.44$

$MSF = SSF/(J - 1) = 3.56/(5 - 1) = .89$

$MSB = SSB/(K - 1) = 2/(3 - 1) = 1$

$MSE = SSE/(J - 1)(K - 1) = 58.44/(5 - 1)(3 - 1) = 7.31$

$MSF/MSE = .89/7.31 = .122 < F(4, 8) = 2.81$

$MSB/MSE = 1/7.31 = .137 < F(2, 8) = 3.11$

Accept the null hypothesis that a salesperson's experience or aptitude test score does not make a difference in sales.

Example 3 **Goodness of Fit Test**

Four hundred students were randomly sampled and asked who they will vote for in the coming student government election. The results are given below.

Candidate	Smith	Gomez	Blackwell	Friedman
Votes	131	121	99	49

Can you argue that the four candidates command different levels of support? Use a 5% level of significance.

Solution: H_0: no difference in support
H_1: different support

$\chi^2 \sim \Sigma (f_o - f_e)^2/f_e$

f_o	f_e	$(f_o - f_e)^2/f_e$
131	100	9.61
121	100	4.41
99	100	0.01
49	100	26.01
	Sum	40.04

$40.04 > \chi^2_{5\%,3} = 7.81$

So we reject the null hypothesis that there is no difference in the support for the four candidates.

Example 4 **Goodness of Fit Test**

Continue the above question. Assume that two days before the election, a debate was held among the four candidates. After the debate, another poll was conducted to study whether the voting pattern changed. Can you argue that the voting pattern changed? Use a 5% level of significance. Below are the results of the new poll.

Candidate	Smith	Gomez	Blackwell	Friedman
Votes	98	102	81	119

Solution: H_0: voting pattern did not change
H_1: voting pattern changed

$\chi^2 \sim \Sigma (f_o - f_e)^2/f_e$

f_o	f_e	$(f_o - f_e)^2/f_e$
98	131	8.31
102	121	2.98
81	99	3.27
119	49	100
	Sum	114.56

$114.56 > \chi^2_{5\%,3} = 7.81$

Reject the null hypothesis that there has been no change in the voting pattern.

Example 5 **Goodness of Fit Tests**

A management consultant wants to develop an inventory control system. She collects demand data for a high tech machine. She wants to determine whether the Poisson distribution fits the data well. Use a 5% level of significance to test whether the data follows a Poisson distribution.

Units Demanded per week	(1) f_o	(2) $P_r(x)$	(3) f_e	(4) $(f_o - f_e)^2/f_e$
0	17	0.05	10	4.9
1	25	0.15	30	0.83
2	40	0.22	44	0.37
3	46	0.22	44	0.09
4	29	0.17	34	0.74
5	28	0.10	20	3.2
6	8	0.05	10	0.4
7	3	0.02	4	0.25
8	4	0.01	2	2
9	0	0.01	2	2
Sum	200	1.00	200	14.77

Solution: To conduct this test, we compare the observed frequency with the expected frequency generated by the poisson distribution:

$$P_r(x) = e^{-\lambda}\lambda^x/x!$$

where λ is the expected number of sales per week. Using this information we can generate the probability of selling x units in column (2). The f_e is obtained as $P_r(x) \times 200$. We then compute

$$\Sigma(f_o - f_e)^2 / f_e = 14.77 < \chi^2_8 = 15.5$$

Therefore, we cannot reject the null hypothesis.

Example 6 **Goodness of Fit Test for Normal Distribution**

Suppose the management consultant of the last example also wants to analyze the sales of a liquid chemical. The sales data are reported below. In her statistical analysis, she assumes that the sales follow a normal distribution. Can you reject the null hypothesis that the data follow a normal distribution at the 5% level of significance?

Sales (gallons)	f_o	f_e	$(f_o - f_e)^2/f_e$
36-38	6	.645	44.46
38-40	22	5.355	51.74
40-42	46	69	7.67
42-44	48	69	6.39
44-46	23	5.355	58.14
46-48	5	.645	29.40
Sum	150		197.8

Solution: The mean and standard deviation should be estimated using the methods for obtaining the group mean and the group standard deviation in Chapter 4.

$$\bar{x} = \sum f \frac{x}{n} = 42$$

$$\text{std.dev} = [\sum f(x - \bar{x})^2 / (n-1)]^{1/2} = 2.28$$

Therefore, we are testing whether the distribution is normal with a mean of 42 and a standard deviation of 2.28. Assuming the null hypothesis is true, we may then obtain the probability of each sales category. For example, the probability of selling more than 46 gallons of the chemical when the sales follows a normal distribution with mean of 42 and standard deviation of 2.28 is obtained as

$$P_r(X > 48) = P_r[(x - 42)/2.28 > (48 - 42)/2.28] = P_r(z > 2.63) = .0043$$

Therefore, the expected sales is $150 \times .0043 = .645$. By using the same process, we may obtain the expected frequency for every category and compute

$$\Sigma(f_o - f_e)^2/f_e = 197.8 > 11.07$$

So we reject the null hypothesis that the data are normally distributed.

Example 7 **Test of Independence**

Three hundred students from the campus are asked who they will vote for in the coming student government election. Their opinions are summarized in the following table.

	Smith	Gomez	Blackwell
Freshman	40	30	60
Sophomore	60	50	80
Junior	80	40	60
Senior	80	30	50

Can you argue that the students' preferences for candidates are independent of their class year? Use a 5% level of significance.

Solution: To answer this question, we need to calculate the expected frequency when the null hypothesis is true. This can be done by using a contingency table.

	Smith		Gomez		Blackwell		Subtotal
	f_o	f_e	f_o	f_e	f_o	f_e	
Freshman	45	40	35	40	80	80	160
Sophomore	25	25	30	25	45	50	100
Junior	50	45	50	45	80	90	180
Senior	30	40	35	40	95	80	160
	150		150		300		600

f_o	f_e	$(f_o - f_e)^2/f_e$
45	40	.625
25	25	0
50	45	.556
30	40	2.5
35	40	.625
30	25	1
50	45	.556
35	40	.625
80	80	0
45	50	.5
80	90	1.111
95	80	2.813
sum		10.91

$$\Sigma(f_o - f_e)^2 = 10.91 < \chi_6^2 = 12.59$$

Therefore, we cannot reject the null hypothesis.

Supplementary Exercises

Multiple Choice

1. Which of the following may be the reason that we can't reject the null hypothesis of $\mu_1 = \mu_2 = \mu_3$ in an ANOVA?
   a. too few groups
   b. too much variation in each group
   c. The degrees of freedom of the numerator is too low.
   d. The degrees of freedom of the denominator is too high.
   e. none of the above

2. The formula for a goodness of fit test is
   a. $(f_o - f_e)/f_e$
   b. $(f_o - f_e)^2/f_o$
   c. $(f_o - f_e)^2/f_e$
   d. $(f_o - f_e)/f_o$
   e. $(f_o + f_e)/f_o$

3. Which of the following is derived assuming the null hypothesis is true?
   a. the expected frequencies
   b. the observed frequencies
   c. the sum of the expected frequencies
   d. the sum of the observed frequencies
   e. the sum of the squared observed frequencies

4. To test whether the data are generated from a binomial distribution you should compute the
   a. expected frequencies assuming the data are not generated from a binomial distribution.
   b. observed frequencies assuming the data are not generated from a binomial distribution.
   c. expected frequencies assuming the data are generated from a binomial distribution.
   d. observed frequencies assuming the data are generated from a binomial distribution.
   e. squared observed frequencies assuming the data are generated from a binomial distribution.

5. Which of the following can be used in a test of independence?
   a. SSB and SSW
   b. SSB, SSW, and SSF
   c. TSS and SSB
   d. ANOVA
   e. contingency table

6. In a one-way ANOVA, a large SSB relative to SSW would most likely lead to
   a. rejection of the null hypothesis of equal means.
   b. acceptance of the null hypothesis of equal means.
   c. rejection of the null hypothesis that the data come from the hypothesized distribution.
   d. acceptance of the null hypothesis that the data come from the hypothesized distribution.
   e. none of the above.

7. In a one-way ANOVA, if there is a large difference between the means of the populations, then
   a. SSB will be large relative to SSW.
   b. TSS will be small.
   c. SSW will be large relative to SSB.
   d. SSW and SSB will be equal.
   e. SSF will be large relative to SSW.

8. ANOVA uses the
   a. t-statistic.
   b. chi-square statistic.
   c. z-statistic.
   d. F-statistic.
   e. binomial distribution.

9. ANOVA is used to test the equality of different population
   a. variances.
   b. means.
   c. modes.
   d. medians.
   e. distributions.

10. In a goodness of fit test, we use the
    a. t-statistic.
    b. chi-square statistic.
    c. z-statistic.
    d. F-statistic.
    e. binomial distribution.

11. In a goodness of fit test, if the expected and observed frequencies are close to one another then most likely
    a. the null hypothesis would be rejected.
    b. the null hypothesis would be accepted.
    c. the alternative hypothesis would be accepted.
    d. the null hypothesis may be accepted or rejected.
    e. both a and c.

12. A contingency table can be used to test
    a. the equality of population means.
    b. the equality of population variances.
    c. statistical independence.
    d. both equality of means and variances.
    e. two-way ANOVA problems.

13. When using a contingency table, the degrees of freedom for the chi-square statistic is equal to the number of
    a. rows.
    b. columns.
    c. rows multiplied by the number of columns.
    d. rows multiplied by the number of columns minus 2.
    e. rows minus 1 multiplied by the number of columns minus 1.

14. In a goodness of fit test, the degrees of freedom for the chi-square statistic is the number of
    a. observations.
    b. observations minus 1.
    c. groups.
    d. groups minus 1.
    e. observations multiplied by the number of groups.

15. In a contingency table, you calculate the expected cell frequency as the
    a. number of observations.
    b. number of groups.
    c. product of row totals times column totals.
    d. product of row totals times column totals divided by sample size.
    e. number of observations divided by sample size.

True/False (if false, explain why)

1. The ANOVA can be used to test the equality of different group means.
2. The idea of a goodness of fit test is to compare the observed frequencies with the expected frequencies, assuming the alternative hypothesis is true.
3. Goodness of fit tests use frequency data.
4. ANOVA usually uses frequency data.
5. If the treatments do not create different effects on the group means, the numerator in the ANOVA will be large.
6. Controlling a second factor in an ANOVA will reduce the sum of squared error.
7. Controlling a second factor in an ANOVA will change the sum of squares between treatment.
8. The sum of the expected frequencies and the sum of the observed frequencies are the same.
9. ANOVA can be used to test whether data are generated from a normal distribution.

10. ANOVA assumes that the sampled populations are normally distributed but have different variances.
11. In a test of independence, the null hypothesis is that the data are dependent.
12. Two-way ANOVA allows us to compare the equality of means when two factors are assumed to influence the population means.
13. When the probability of one variable occurring is not influenced by the outcome of another variable, the two variables are statistically independent.
14. A contingency table uses an F-statistic to test for statistical independence.
15. In ANOVA, the null hypothesis is that the data come from a single population.

Questions and Problems

1. Four hundred television viewers were asked what T.V. news they usually watched last year. The results were

ABC	140
NBC	100
CBS	100
CNN	60

Can you reject the null hypothesis that the ratings are the same? Use a 5% level of significance.

2. The same four hundred T.V. viewers are asked the same question again a year later. The results are as follows

ABC	100
NBC	120
CBS	120
CNN	60

Can you argue that the rating pattern has changed? Use a 5% level of significance.

3. The following data represent gallons of gasoline sold each in a gas station.

Gallons	Number of Days
< 1400	20
1400-1600	30
1600-1800	50
1800-2000	45
2000-2200	35
> 2200	20
	200

Do a 5% test to determine if there is enough evidence to show that the data do not come from a normal distribution with a mean of 1800 and a standard deviation of 200.

4. Six hundred economists were asked about their opinions regarding the new budget. Their background and opinions are summarized in the following table.

Background	Opinion on Budget			
	like it	indifferent	hate it	
conservative	100	80	60	240
liberal	80	60	40	180
radical	60	50	70	180
	240	190	170	600

Do a 5% test to determine if the economists' opinions depend on their background.

5. Use the data from Question 4 to determine if radical economists' opinions are equally split. Use a 5% test.

6. Use the data from Question 4 to determine if the pattern of reactions to the budget is the same for liberal and radical economists. Use a 5% test.

Answers to Supplementary Exercises

Multiple Choice

1. b	6. a	11. b
2. c	7. a	12. c
3. a	8. d	13. e
4. c	9. b	14. d
5. e	10. b	15. d

True/False

1. True
2. False. Compare the observed frequencies with the expected frequencies assuming the null hypothesis is true.
3. True
4. False. ANOVA uses between group and within group variances.
5. False. The numerator will be small.
6. True
7. False. It does not change.
8. True
9. False. Goodness of fit tests are used to test if data come from a normal distribution.
10. False. ANOVA assumes equal population variances.
11. False. The null hypothesis is that the data are independent.
12. True
13. True
14. False. Contingency table uses a chi-square statistic.
15. True

Questions and Problems

1. If we are interested in testing that the ratings are the same for each station, we are assuming that the ratings are uniformly distributed. For a uniform distribution, the expected frequency would be the total viewers surveyed divided by the number of groups, 400/4 = 100.

Network	f_o	f_e	$(f_o - f_e)^2/f_e$
ABC	140	100	16
NBC	100	100	0
CBS	100	100	0
CNN	60	100	0
Sum		$\chi^2 =$	32

$\chi^2_{3,.05} = 7.185$

Reject the null hypothesis that the ratings are equal.

2. In this case, we want to test the current ratings (observed frequency) against the past ratings (expected frequency).

Network	f_o	f_e	$(f_o - f_e)^2/f_e$
ABC	100	140	11.43
NBC	120	100	4
CBS	120	100	4
CNN	60	60	0
Sum		$\chi^2 =$	19.43

Reject the null hypothesis that the ratings have not changed.

3. To solve this question, we need to compute the expected frequencies.

$$P_r(1400 > \text{gallons sold}) = P_r\left(\frac{1400-1800}{200} > \frac{\text{gallons}-1800}{200}\right)$$
$$= P_r(-2 > z) = 2.28\%$$

$$f_e = 200 \times .028 \;=\; 4.56$$

We can compute the other expected frequencies similarly.

Gallons Sold	f_o	f_e	$(f_o - f_e)^2/f_e$
< 1400	20	4.56	52.28
1400-1600	30	27.18	0.29
1600-1800	50	67.87	4.71
1800-2000	45	67.87	27.18
2000-2200	35	27.18	2.25
> 2200	20	4.56	52.28
	200	$\chi^2 =$	119.51

$$\Sigma \frac{(f_o - f_e)^2}{f_e} = 119.51 > \chi_{.05,5} = 11.07$$

So we reject the null hypothesis that the data are distributed normally with mean of 1800 and standard deviation of 200.

4. To test for the statistical independence of the economists' background and their opinion, we need to calculate the expected frequencies. The expected frequency for each cell is the product of the column total times the row total divided by the total number of economists.

Expected Frequencies				
conservative	96	76	68	240
liberal	72	57	51	180
radical	72	57	51	180
	240	190	170	600

f_o	f_e	$(f_o - f_e)^2/f_e$
100	96	.167
80	76	.210
60	68	.941
80	72	.889
60	72	.158
40	51	1.588
60	72	.889
50	57	.860
70	51	7.078
Sum	$\chi^2 =$	12.78

$\chi_{,.05} = 9.488$, so reject the null hypothesis that the economists' opinions are independent of their background.

5. We are assuming a uniform distribution with the expected frequency equal to 60.

f_o	f_e	$(f_o - f_e)^2/f_e$
60	60	0
50	60	1.667
70	60	1.667
Sum	$\chi^2 =$	3.333

$\chi^2_{2,.05} = 5.991$

so we cannot reject the null hypothesis that the opinions of the radical economists are uniformly distributed.

6. H_0: same pattern vs. H_1: different pattern

f_o	f_e	$(f_o - f_e)^2/f_e$
80	60	6.67
60	50	2
40	70	12.86
Sum	$\chi^2 =$	21.52

$\chi^2_{2,.05} = 5.991$

so reject the null hypothesis.

CHAPTER 13

SIMPLE LINEAR REGRESSION AND THE CORRELATION COEFFICIENT

Chapter Intuition

We are often interested in measuring the relationship between two variables. In some cases, we know that the relationship is linear, and so we can use linear regression analysis to measure it. For example, elementary economics textbooks state that consumption is a function of income. The equation for the consumption function is

$$\text{CONSUMPTION} = \alpha + \beta\ [\text{INCOME}]$$

The equation for the consumption function represents a straight line, with intercept of α and slope of β. Besides knowing that the consumption function is a straight line, we know that the slope coefficient, β, should be positive, because consumption should rise as income increases.

If we are interested in the actual values of α and β, and hence, the actual equation that describes the relation between consumption and income, then we will collect data and use regression analysis to estimate this relationship.

Simply stated, regression analysis tries to fit a line through a scatter plot of points of the dependent and independent variables. Although there are many ways to fit the line through these points, the most popular approach is the ***least squares method***. The least squares method fits the regression line by minimizing the sum-of-squared error terms. The slope β of the line measures the impact of x (the independent variable) on y (the dependent variable). The intercept term, α, measures the value of y when the value of x is zero.

Once we have determined the regression line by estimating a and b, we are interested in how good a job the line does of explaining the relationship between the two variables. If most of the data points are close to the regression line, then the sum-of-squared error terms will be small and the regression line will explain much of the relationship between the two variables. Likewise, if the data points tend to be far away from the regression line, the sum-of-squared error terms will be large and the regression line will not do a good job of explaining the relationship. This relationship can be seen by looking at the regression's coefficient of determination, or R^2. The coefficient of determination measures how much of the variance of the dependent variable can be explained by the regression.

Chapter Review

1. ***Correlation analysis*** is used to measure the relationship between two different variables. The correlation coefficient, ρ, between two variables will lie between –1, perfect negative correlation and +1, perfect positive correlation. If ρ is positive, the two variables tend to move in the same direction (when one goes up, the other usually goes up). If ρ is negative, the two variables tend to move in opposite directions (when one goes up, the other usually goes down).
2. ***Simple linear regression*** is a systematic procedure for finding a linear relationship between two variables.
3. In simple linear regression, there are only two variables of interest. The ***independent variable*** or ***explanatory variable***, x, is the variable which we assume will explain the **dependent variable**, y.
4. In regression analysis, there are two parameters of interest. The **intercept** coefficient represents the value of y if x is zero. The **slope** coefficient, which is usually represented by β, measures the impact of x on y.
5. Sometimes a ***scatter diagram*** is used to depict the relationship between the x and y variables. A scatter diagram is just a plot of the x and y variables. Geometrically, regression analysis is just a systematic way of fitting a line through the scatter diagram to best describe the relationship between the x and y variables.
6. One method used for fitting a line through the scatter diagram is known as the ***least squares method.*** The least squares method fits the line through the scatter diagram by minimizing the total-squared error terms from the regression line.

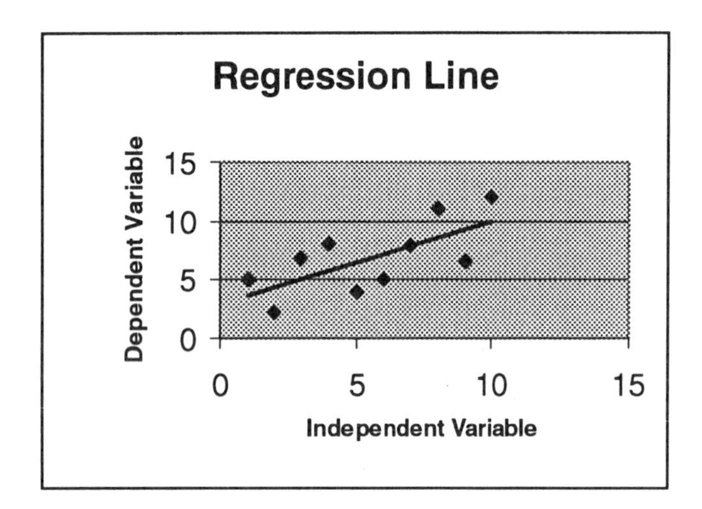

7. The assumptions for the linear regression model are:

   a. The values of the independent variable, x, are either fixed numbers or if they are random variables, they are statistically independent of the error term, ε_i.

   b. The error terms, ε_i, are assumed to have a mean of 0 and a constant variance. They are also assumed to be statistically independent of one another.

8. Once we have estimated the regression line, we would like to know how well this line represents the relationship between x and y. To measure the goodness of fit of the regression line, we can use the **standard error of the residuals** or the **coefficient of determination.** The standard error of the residuals measures the variability of the observed values around the regression line. The coefficient of determination tells us how much of the variation of y is explained by x.

Useful Formulas

Basic regression equation:

$$y = \alpha + \beta x + \varepsilon$$

Least squares estimates:

$$b = \frac{\sum_{i=1}^{n}(x_i - \bar{x})(y_i - \bar{y})}{\sum_{i=1}^{n}(x_i - \bar{x})^2}$$

$$a = \bar{y} - b\bar{x}$$

Sum-of-squares regression:

$$SSR=\Sigma(\hat{y}_i-\bar{y})^2$$

Sum-of-squares residual:

$$SSE=\Sigma(y_i-\hat{y}_i)^2$$

Sum-of-squares total:

$$SST=\Sigma(y_i-\bar{y})^2$$

$$SST = SSR + SSE$$

Coefficient of determination:

$$R^2 = \frac{SSR}{SST} = 1-\frac{SSE}{SST}$$

Adjusted coefficient of determination:

$$\overline{R}^2 = 1-\frac{SSE/(n-k-1)}{SST/(n-1)}$$

Standard error of the residual:

$$s_e = \sqrt{\frac{\sum_{i=1}^{n}(y_i-\hat{y}_i)^2}{(n-k-1)}} = \sqrt{\frac{SSE}{(n-k-1)}}$$

Correlation coefficient:

$$\rho = \frac{Cov(x,\ y)}{\sigma_x\ \sigma_y}$$

Example Problems

Example 1 **Regression Line**

Suppose you collect data on household consumption and income in the U.S. and estimate the following regression equation:

$$C = 1{,}200 + .75\ Y$$

where C = consumption
Y = income

a. What is the dependent variable? What is the independent variable?

b. Plot the relationship between consumption and income from the above equation.

c. Explain the relationship between consumption and income. How much more will consumption increase if income rises by $1?

Solution:

a. Because consumption depends on income, consumption is the dependent variable, while income is the independent variable.

b.

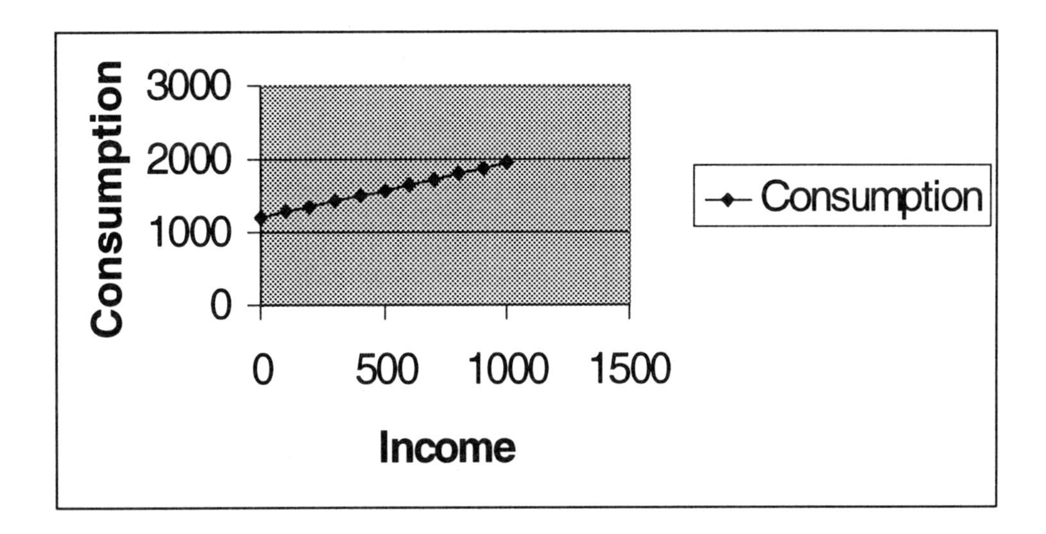

c. The above equation for the consumption function says that consumption has two parts. The first part is represented by the intercept term of 1,200. This intercept term means that consumption will be 1,200 even if there is no income. The second part consists of the .75 Y term. This says that for every $1 increase in income (Y), consumption will increase by 75 cents.

Example 2 **Scatter Diagram**

Suppose you are a quality control consultant for the Hug the Road Tire Company. You are interested in the relationship between the number of miles a tire has and the number of flat tires. You collect the following information for six different tires.

Tire	Miles driven	# flats
1	10,000	0
2	11,900	1
3	27,000	2
4	35,000	3
5	50,000	2
6	42,000	3

a. Draw a scatter diagram showing the relationship between miles on the tire and the number of flat tires.

b. Is there a direct or an inverse relationship between miles driven and the number of flats?

Solution:

a.

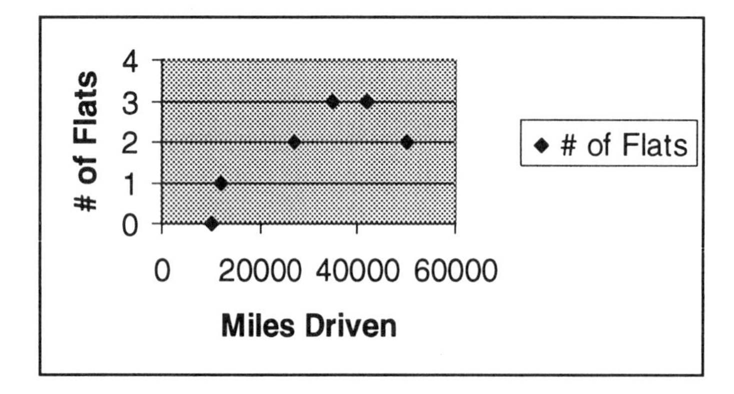

b. There appears to be a direct relationship; that is, as the number of miles increases, the number of flats also increases.

Example 3 **Estimating Regression Coefficients**

Suppose you collect data for Orange Computer's sales and dollars spent on research and development. You compute the following statistics:

Cov(sales, R&D) = 300
Var(sales) = 125
Var(R&D) = 880
Mean sales = 1,200
Mean R&D = 895

a. Compute the correlation coefficient between R&D and sales.

b. Calculate the coefficient of determination which would result from a regression of sales on R&D.

c. Calculate the regression parameters *a* and *b* .

Solution:

a.

$$r_{x,y} = \frac{\text{Cov}(x,y)}{\sigma_x \sigma_y} = \frac{300}{\sqrt{125}\ \sqrt{880}} = .286$$

b.

$$R^2 = r^2 = .286^2 = .081$$

c.

$$b = \frac{\text{Cov(sales, R\&D)}}{\text{Var(R\&D)}} = \frac{300}{1250} = .24$$

$$a = \text{mean sales} - b\,(\text{mean R\&D}) = 1200 - .24\,(895) = 985.2$$

Example 4 **Coefficient of Determination**

Suppose you estimate a regression and compute SSE and SSR as

SSE = 53.27
SSR = 202.91

Calculate SST, R^2 and r using the above information.

Solution:

SST = SSE + SSR = 53.27 + 202.91 = 256.18

R^2 = SSR/SST = 202.91/256.18 = .792

r = $\sqrt{.792}$ = .890

Example 5 **Estimating Regression Coefficients**

You are given the following table

x	y	$(x_i - \bar{x})$	$(y_i - \bar{y})$	$(x_i - \bar{x})(y_i - \bar{y})$	$(x_i - \bar{x})^2$
20	75				
14	85				
12	92				
20	88				
33	72				
38	99				

Fill in the missing values and solve for the regression coefficients a and b.

Solution:

	x	y	$(x_i - \bar{x})$	$(y_i - \bar{y})$	$(x_i - \bar{x})(y_i - \bar{y})$	$(x_i - \bar{x})^2$
	20	75	–2.83	–10.17	28.81	8.03
	14	85	–8.83	–0.17	1.47	78.03
	12	92	–10.83	6.83	–74.03	117.36
	20	88	–2.83	2.83	–8.03	8.03
	33	72	10.17	–13.17	–133.86	103.36
	38	99	15.17	13.83	209.81	230.03
Sum	137	511			24.17	544.84
Mean	22.83	85.16				

$$b = \frac{\Sigma(x_i - \bar{x})(y_i - \bar{y})}{\Sigma(x_i - \bar{x})^2} = .04436$$

$$a = \bar{y} - b\bar{x} = 85.16 - .04436(22.83) = 84.15$$

Supplementary Exercises

Multiple Choice

1. A scatter diagram is
   a. a line that measures the relationship between the independent and dependent variables.
   b. another name for simple regression.
   c. another name for multiple regression.
   d. a graph of values of the independent and dependent variables.
   e. a line with slope of b and intercept of a.

2. A regression line
   a. is a line that measures the relationship between the independent and dependent variables.
   b. is another name for a scatter diagram.
   c. always has a slope equal to 1.
   d. is a graph of values of the independent and dependent variables.
   e. always has a slope equal to 0.

3. The slope coefficient
   a. is the point where the regression line intersects the *y*-axis.
   b. measures the fit of the regression line.
   c. measures the relationship between the independent and dependent variables.
   d. is always equal to 1.
   e. is another name for the coefficient of determination.

4. The intercept term
   a. is the point where the regression line intersects the y-axis.
   b. measures the fit of the regression line.
   c. measures the relationship between the independent and dependent variables.
   d. is always equal to 1.
   e. is always equal to 0.

5. Which of the following is *not* an assumption of the linear regression model?
   a. Either x_i are fixed numbers or they are statistically independent of the random variables ε_i.
   b. The variance of the random variable ε_i is assumed to be constant.
   c. The random variable ε_i is assumed to have a mean of 0.
   d. The variance of the random variable ε_i is assumed to be 0.
   e. The random variables ε_i are assumed to be statistically independent of one another.

6. The coefficient of determination
   a. is the point where the regression line intersects the *y*-axis.
   b. measures the fit of the regression line.
   c. measures the relationship between the independent and dependent variables.
   d. is always equal to 1.
   e. is always equal to 0.

7. The standard error of the estimate
   a. is the point where the regression line intersects the *y*-axis.
   b. measures the fit of the regression line.
   c. measures the relationship between the independent and dependent variables.
   d. is always equal to 1.
   e. is another name for the coefficient of determination.

8. The coefficient of determination measures the
   a. variation of the independent variable.
   b. slope of the regression line.
   c. intercept of the regression line.
   d. total variation of the dependent variable that is explained by the regression.
   e. is always equal to 1.

9. SST is
   a. $\sum(y_i - \bar{y})^2$
   b. $\sum(\hat{y}_i - \bar{y})^2$
   c. $\sum(y_i - \hat{y})^2$
   d. SSR – SSE
   e. SSE – SSR

10. SSR is
   a. $\sum(y_i - \bar{y})^2$
   b. $\sum(\hat{y}_i - \bar{y})^2$
   c. $\sum(y_i - \hat{y})^2$
   d. SSE + SST
   e. SSE – SST

11. SSE is
   a. $\sum(y_i - \bar{y})^2$
   b. $\sum(\hat{y}_i - \bar{y})^2$
   c. $\sum(y_i - \hat{y})^2$
   d. SSE – SST
   e. SSR + SST

12. The covariance between x and y is
   a. $\sum(x_i - \bar{x})^2$
   b. $\sum(y_i - \bar{y})^2$
   c. $(1/n)\sum(x_i - \bar{x})(y_i - \bar{y})$
   d. $r_{xy}\,\sigma_x\sigma_y$
   e. both c and d.

13. If we were interested in measuring the relationship between work experience and earnings using regression analysis,
   a. the independent variable should be earnings.
   b. the independent variable should be work experience.
   c. the dependent variable should be earnings.
   d. the dependent variable should be work experience.
   e. both b and c.

14. If you estimate the following regression, y = .34 + 1.2 x, then the slope coefficient is
   a. x
   b. y
   c. .34
   d. 1.2
   e. 1.2/.34

15. If you estimate the following regression, *y* = .34 + 1.2 x, then the intercept term is
   a. x
   b. y
   c. .34
   d. 1.2
   e. 1.2/.34

16. If you estimate the relationship between earnings (in dollars) and education (in years) as EARN = 12,201 + 525 EDUC, then a person with 2 additional years of education would be expected to earn an additional
    a. $12,201
    b. $525
    c. $24,402
    d. $1,050
    e. $12,201 + $525

17. If you estimate the relationship between earnings (in dollars) and education (in years) as EARN = 12,201 + 525 EDUC, then a person with zero years of education would be expected to earn
    a. $12,201
    b. $525
    c. $24,402
    d. $1,050
    e. $12,201 + $525

18. If a regression has an R^2 of .80, then the regression line
    a. explains 80% of the variation in x.
    b. explains 80% of the variation in y.
    c. will have a slope of .80.
    d. will have an intercept term of .80.
    e. will not do a good job of explaining the relationship between x and y.

True/False (If false, explain why)

1. A scatter diagram measures the relationship between the dependent and independent variables.
2. In regression analysis, the dependent variable is usually placed on the x-axis and the independent variable is usually placed on the y-axis.
3. If we are only interested in measuring the relationship between two variables, correlation analysis can be as useful as regression analysis.
4. The fit of the regression model can be measured by using either the coefficient of determination or the standard error of the estimate.
5. Multiple regression is used when the independent variable is influenced by two or more dependent variables.
6. The slope coefficient in a regression measures the relationship between the dependent and independent variables.
7. The intercept term in a regression is an estimate of the value of the dependent variable when the independent variable is 0.

8. If we are interested in using regression analysis to measure the relationship between high school and college grades, the dependent variable should be high school grades and the independent variable should be college grades.
9. In correlation analysis, it is important to know which is the independent variable and which is the dependent variable.
10. The method of least squares fits the regression line by minimizing the sum of the squared residuals.
11. The coefficient of determination measures the percentage of the independent variable's variance that is explained by the regression.
12. If we are interested in comparing the relationship between two pairs of variables, A and B, and X and Y, we can use either covariance or correlation.
13. If two variables have a correlation coefficient of –1, they will always move in opposite directions.
14. If an economist reverses the roles of the explanatory variable and the explained variable, the R^2 will decrease.
15. The R^2 of a simple regression is equal to the square of the correlation coefficient between x and y.
16. If $b \leq 0$, this implies that x has little impact on y.
17. If the correlation coefficient equals 1, the data points lie on a straight line.
18. Covariance will always lie between –1 and +1.

Questions and Problems

1. You are given the following information from a regression:

   SSE = 28.6 SSR = 30.1 n = 50

   a. Compute the coefficient of determination.

   b. Compute the standard error of the estimate.

2. You are given the following information about two variables, x and y:

   $\overline{x} = 32 \quad \overline{y} = 6 \quad Cov(x, y) = 81 \quad Var(x) = 21$

   Compute the parameters for the slope and intercept of a regression of y on x.

3. Suppose you are interested in the relationship between income and years of schooling. You estimate the following regression:

   $INCOME_i = 12{,}000 + 1{,}250\ SCHOOLING_i$

   If income is measured in dollars and schooling is measured in years, interpret the results of this regression.

4. Compute the correlation coefficient between *x* and *y*, given the following information.

Observation	x	y
1	10	22
2	9	31
3	11	19
4	6	25

5. Use the information given in Problem 4 to plot a scatter diagram.
6. Use the information in Problem 4 to estimate the regression of y on x.
7. Suppose you estimate the consumption function as

   CONS = 1,225 + .82 INCOME

   How much is consumption expected to increase if income rises by $1,000? How much will consumption be if income is $0?

8. Suppose the R^2 from the regression in Problem 7 is .75. Briefly explain what this means.

Answers to Supplementary Exercises

Multiple Choice

1. d	6. b	11. c	16. d
2. a	7. b	12. e	17. a
3. c	8. d	13. e	18. b
4. a	9. a	14. d	
5. d	10. b	15. c	

True/False

1. False. A scatter diagram is a plot of the independent and dependent variables.
2. False. x-independent variable and y-dependent variable.
3. True
4. True
5. False. Multiple regression is used when there are two or more independent variables.
6. True
7. True
8. False. High school grades would be the independent variable and college grades would be the dependent variable.
9. False. In correlation analysis, it doesn't matter which is the dependent variable and which is the independent variable.
10. True
11. False. It measures the percentage of the dependent variable's variance that is explained by the regression.
12. False. We need to use correlation to compare the degree of association between two pairs of variables because correlation is unit-free.
13. True
14. False. R^2 remains the same.
15. True
16. False. It may have significant negative impact.
17. True
18. False. Correlation will lie between –1 and +1; covariance can take on any value.

Questions and Problems

1. a. $R^2 = \frac{SSR}{SST} = \frac{30.1}{28.6 + 30.1} = .513$

 b. $s_e = \sqrt{\frac{SSE}{n-2}} = \sqrt{\frac{28.6}{50-2}} =.772$

2. $b = \frac{Cov(x,y)}{Var(x)} = \frac{81}{21} = 3.86$

 $a = \bar{y} - b\bar{x} = 6 - 3.86(32) = -117.52$

3. The intercept term says that a person with no schooling would be expected to earn $12,000. The slope coefficient says that for each additional year of schooling, income is expected to rise by $1,250.

4.

	x	y	$(x-\bar{x})(y-\bar{y})$	$(x-\bar{x})^2$	$(y-\bar{y})^2$
	10	22	–2.25	1	5.06
	9	31	0	0	45.56
	11	19	–10.5	4	27.56
	6	25	–2.25	9	0.56
Mean	9	24.2			
Sum			–15	14	78.75

$$r_{x,y} = \frac{-15}{\sqrt{14}\,\sqrt{78.75}} = -0.452$$

5.

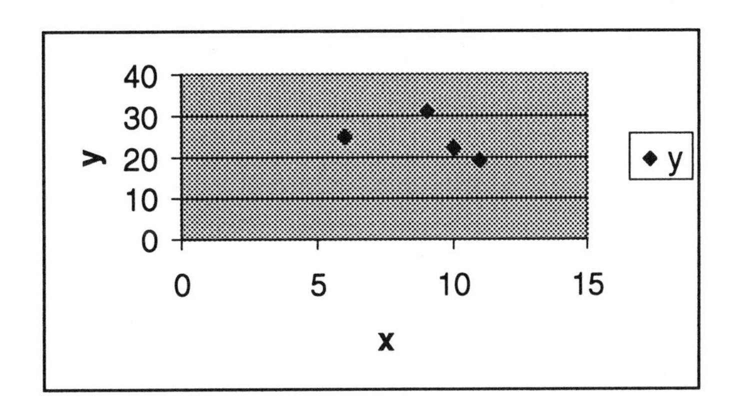

6. $b = \dfrac{-15}{14} = -1.07$

$a = 24.2 - (-1.07)(9) = 33.83$

7. An increase of $1,000 in income is expected to lead to an $820 increase in consumption. If income is $0, then consumption would be $1,225.

8. An R^2 of .75 means that 75% of the variation in consumption is explained by the regression of consumption on income.

CHAPTER 14

SIMPLE LINEAR REGRESSION AND CORRELATION: ANALYSES AND APPLICATIONS

Chapter Intuition

In Chapter 13, you learned how regression analysis and correlation can be used to find a relationship between two variables. The regression line that you learned about provided an estimate of the relationship between the explanatory variables and the explained (or response) variables. However, as you learned in Chapter 10, estimates are nothing more than educated guesses about the true value. Because we are not absolutely certain that our estimates are meaningful, we use statistical theory to test the significance (or reliability) of the parameters we estimated. We can also use statistical theory to provide upper and lower bounds (known as a confidence interval) for the true values of the slope and intercept terms.

When we test the significance of the slope coefficient, we are interested in seeing if the explanatory variable, x, impacts the value of the y variable. If the slope coefficient is not statistically different from 0, then the explanatory variable is not useful in explaining values of the dependent variable, y.

Chapter Review

1. Once we have estimated the regression coefficients, we are interested in testing the significance of these coefficients. Generally, we test to see if the coefficients are significantly different from zero, although in some instances, we may be interested in seeing if the slope coefficient is different from some other number.

2. If the error terms, ε_i, follows a normal distribution, then b will follow a t-distribution with n – 2 degrees of freedom. Statistical inference can be done on the slope coefficient, b.

3. Testing the significance of the slope in a simple regression can be done by using a t-test.

4. In using regression analysis to ***forecast*** the values of y, we are sometimes interested in constructing an upper and lower band around the forecast. This band is known as a **confidence belt** (or confidence interval) around the forecast. This is done because we are uncertain of the true value of the forecasted y, and therefore would like to provide some upper and lower bounds on the likely value of the forecasted y. This confidence interval is not represented by parallel lines around the regression line, but rather by curves that get farther away from the regression line as we get farther away from the mean of x. This is because the farther we are away from the center of our data, , the less certain we are about the value of our forecasted y.

5. One of the assumptions of the simple linear regression model is that the independent variable, x, is measured without error. However, in economics and business, very often the x variable is measured with error. This results in an errors-in-variable problem. The result of measurement error in the x variable is that the estimator of the slope parameter will be biased.

6. The ***market model*** represents one of the most commonly used regressions in finance. It is used to measure the relationship between the returns on a company's stock and the returns on some market index like the DJIA or S&P 500. The estimated slope coefficient is one of the most commonly used measures of a stock's risk.

Useful Formulas

Significance test for $b = b_0$:

$$t_{n-2} = \frac{b - b_0}{s_b}$$

$$s_b = \frac{s_e}{\sqrt{\Sigma(x-\bar{x})^2}}, \quad s_e = \sqrt{\frac{\Sigma e^2}{n-2}}$$

Significance test for $a \neq 0$:

$$t_{n-2} = \frac{a - 0}{s_a}$$

$$s_a = \sqrt{s_e^2 \frac{\Sigma x^2}{n\Sigma(x-\bar{x})^2}}$$

F-test for the significance of b:

$$F_{1,n-2} = \frac{SSR/1}{SSE/(n-2)} = \frac{MSR}{MSE}$$

Significance test for r:

$$t_{n-2} = \frac{r\sqrt{n-2}}{\sqrt{1-r^2}}$$

Conditional expectation interval:

$$\hat{y}_{n+1} \pm t_{\alpha/2,n-2} s_e \sqrt{\frac{1}{n} + \frac{(x_{n+1} - \bar{x})^2}{\sum_{i=1}^{n}(x_i - \bar{x})^2}}$$

Prediction interval:

$$\hat{y}_{n+1} \pm t_{\alpha/2,n-2} s_e \sqrt{1 + \frac{1}{n} + \frac{(x_{n+1} - \bar{x})^2}{\sum_{i=1}^{n}(x_i - \bar{x})^2}}$$

Market model:

$$R_{jt} = a + bR_{mt} + e_{jt}$$

Example Problems

Example 1 **t-tests for Parameter Estimates**

When estimating the relationship between the price of a good and quantity of the good sold (the demand curve), economists sometimes choose to transform the price and quantity data by taking the natural logarithm of both. When this is done, the slope coefficient β can be interpreted as the price elasticity of demand (the sensitivity of quantity demanded to changes in price). Below you are given information on the price and quantity sold of Thirsty Time Cola.

Price	Quantity
$2.20	1020
1.85	1521
1.50	1755
1.21	2130
.99	3200
.79	4105

a. Estimate the elasticity of demand for the above data.

b. Use a *t*-test to test the significance of *b*.

c. Construct a 90% confidence interval for the price elasticity.

Solution:

	p ln(*P*)	*q* ln(*Q*)	$(p-\bar{p})^2$	$(q-\bar{q})^2$	$(p-\bar{p})(q-\bar{q})$	$\hat{q}$	e	e^2
	.788	6.928	.246	.493	–.348	6.986	–.058	.003
	.615	7.327	.104	.092	–.098	7.211	.116	.014
	.406	7.470	.013	.026	–.018	7.483	–.013	.000
	.191	7.664	.010	.001	–.003	7.762	–.098	.010
	–.010	8.071	.019	.194	–.133	8.022	.048	.002
	–.236	8.320	.279	.476	–.364	8.315	.005	.000
Mean	.292	7.630	.124	.214	–.161			
Sum			.744	1.282	-.965	45.780	.029	

$$b=\frac{\Sigma(x-\bar{x})(y-\bar{y})}{\Sigma(x-\bar{x})^2}=\frac{-.965}{.744}=-1.298$$

$$a=\bar{y}-b\bar{x}=7.630-(-1.298)(.292)=8.009$$

$$s_e^2=\frac{SSE}{(n-2)}=\frac{\Sigma e^2}{(n-2)}=\frac{.0291}{(6-2)}=.0073$$

$$s_e=\sqrt{.0073}=.0853$$

$$s_b^2=\frac{s_e^2}{\Sigma(x-\bar{x})^2}=\frac{.0073}{.744}=.0098$$

$$s_b=\sqrt{.0098}=.0989$$

$$b\pm t_{\alpha/2,n-2}s_b$$

$$-1.298\pm 2.776(.0989)$$

$$-1.572 \text{ to } -1.023$$

Example 2 **Confidence Intervals Around Forecasts**

Suppose you use the data in Example 1 to estimate the demand for Thirsty Time Cola without using a log transformation, and you obtain the following results,

$$Q_i = 5149.268 - 2009.90\ P_i$$

$$s_e = 468.943,\ R^2 = .867$$

$$\Sigma(P_i - \hat{P})^2 = 1.426$$

Now suppose the company charges $1.75 for Thirsty Time Cola. Forecast the demand for the cola and construct a 90% confidence interval around the forecast.

Solution:

$$\hat{Q}_{n+1} = 5149.268 - 2009.90(1.75) = 1{,}631.9$$

$$\hat{y}_{n+1} \pm t_{\alpha/2,n-2} s_e \sqrt{\frac{1}{n} + \frac{(x_{n+1} - \bar{x})^2}{\sum_{i=1}^{n}(x_i - \bar{x})^2}}$$

$$1{,}631.9 \pm 2.132(468.943)\sqrt{\frac{1}{4} + \frac{(1.75 - 1.423)^2}{1.426}}$$

So the interval is from 1,061.05 to 2,201.85.

Example 3 **Confidence Interval for Parameters**

Suppose you estimate a regression of y against x and find the following.

$b = 2.8$
$\sigma_b = .6$ standard error of the regression (assumed known)

a. Construct a 95% confidence interval for b.

b. Suppose σ_b is not known, how would this change the way you construct your confidence interval for b? Assume that 25 observations were used to estimate the regression.

Solution:

a. $b \pm z_{\alpha/2}\,\sigma_b$
$2.8 \pm 1.96(.6)$

So the interval is from 1.624 to 3.976.

b. If σ_b is not known, we must use an estimate for σ_b, s_b. In this case, we should use the Student's t distribution to construct our confidence interval.

$b \pm t_{\alpha/2,n-2}\,s_b$
$2.8 \pm 2.069(.6)$

So the interval is from 1.559 to 4.041.

Example 4 **Significance Tests**

Suppose you estimate the following simple regression:

$$y = 2,300 + 10.12\ x$$

$$SSE = 28,225$$
$$n = 28$$
$$\sum(x - \bar{x})^2 = 3,300$$

a. Based on the information provided, test the significance of the slope at the 95% confidence level.

b. Construct a 90% confidence interval for the slope coefficient.

Solution: $s_e^2 = SSE/(n-2) = 28,225/(28-2) = 1085.58$

$$s_e = \sqrt{1085.58} = 32.95$$

$$s_b = s_e / \left[\sum(x - E(x))^2\right]^{1/2} = 32.95/\sqrt{3,300} = .5736$$

a. $t = [b - E(b)]/s_b = [10.12 - 0]/.5736 = 17.64$

Since the critical value is 2.056, we clearly reject the null hypothesis that b = 0 and conclude that b is significant.

b. $b \pm t_{\alpha/2,n-2}\ s_b$

$10.12 \pm 1.706(.5736)$

So the interval is from 9.14 to 11.10.

Example 5 **Joint Test of Significance**

Suppose you estimate the following relationship between inches of rainfall and amount of corn in bushels for 100 farmers:

$$\text{Bushels}_i = 5{,}100 + 1{,}100\ \text{Rainfall}_i$$
$$SSE = 3{,}221$$
$$SSR = 6{,}755$$

Use an *F* test to test the significance of this regression at the 95% level.

Solution:

$$F_{1,n-2} = \frac{SSR/1}{SSE/(n-2)} = \frac{6755/1}{3221/(100-2)} = 205.52$$

The critical value for $F_{1,98,\,5\%}$ is approximately 4.00, so we can clearly conclude that the regression is significant.

Supplementary Exercises

Multiple Choice

1. When examining the significance of the regression coefficient, the null hypothesis is usually
   a. H_0: $\beta = 1$
   b. H_0: $\beta = -1$
   c. H_0: $\beta = 0$
   d. H_0: $\beta = 0.5$
   e. H_0: $\beta = -0.5$

2. For a two-tailed test of the significance of β, the alternative hypothesis is usually
   a. H_1: $\beta \neq 1$
   b. H_1: $\beta > 1$
   c. H_1: $\beta \neq 0$
   d. H_1: $\beta > 0$
   e. H_1: $\beta < 0$

3. If we are interested in examining if the slope coefficient is *positive* and significant, the null hypothesis should be
   a. H_0: $\beta \leq 0$
   b. H_0: $\beta \geq 0$
   c. H_0: $\beta = 0$
   d. H_0: $\beta = 1$
   e. H_0: $\beta \geq 1$

4. If we are interested in examining if the slope coefficient is *negative* and significant, the null hypothesis should be
   a. H_0: $\beta \leq 0$
   b. H_0: $\beta \geq 0$
   c. H_0: $\beta = 0$
   d. H_0: $\beta = 1$
   e. H_0: $\beta = -1$

5. When examining the significance of a regression parameter, we usually use the
   a. F-test.
   b. chi-square test.
   c. t-test.
   d. binomial distribution.
   e. any of the above.

6. When examining the significance of a regression, we are
   a. most interested in examining the significance of the intercept term α.
   b. most interested in examining the significance of the slope coefficient β.
   c. equally interested in the intercept and the slope.
   d. most interested in the variance of y.
   e. most interested in the variance of x.

7. If the slope coefficient β is insignificant, it means that
   a. the independent variable does a good job of explaining the dependent variable.
   b. the independent variable does not do a good job of explaining the dependent variable.
   c. the dependent variable does a good job of explaining the independent variable.
   d. the dependent variable does not do a good job of explaining the independent variable.
   e. none of the above.

8. If the slope coefficient for a regression is 2.4 and the standard error of the slope coefficient is .8, then the *t*-value used to test H_0: $\beta = 0$ is
   a. 8./2.4
   b. $2.4/\sqrt{.8}$
   c. (2.4 – 1)/.8
   d. 2.4/.8
   e. $(2.4-1)/\sqrt{.8}$

9. If the slope coefficient for a regression is 2.4 and the standard error of the slope coefficient is .8, then the *t*-value used to test H_0: $\beta = 1$ is
   a. 8./2.4
   b. $2.4/\sqrt{.8}$
   c. (2.4 – 1)/.8
   d. 2.4/.8
   e. $(2.4-1)/\sqrt{.8}$

10. Other things being equal, the greater the standard error of the slope coefficient, the
    a. larger the *t*-value for the slope coefficient.
    b. smaller the *t*-value for the slope coefficient.
    c. larger the intercept term.
    d. smaller the intercept term.
    e. larger the slope coefficient.

11. In simple regression, a significant *t*-value for the slope coefficient (for H_0: $\beta = 0$), implies that
    a. the F-statistic will be significant for the slope coefficient.
    b. the F-statistic will not be significant for the regression.
    c. the F-statistic could be significant or insignificant.
    d. we cannot compare the results from a t-test and F-test in simple regression.
    e. the R^2 will be very small.

12. Which of the following statements could be true?
    a. $SSE + SSR > SST$
    b. $R^2 = -.5$
    c. $R^2 = 1.83$
    d. $s_e = -.35$
    e. $t = -2.3$

13. In a regression, it is always true that
    a. $r > 0$
    b. $r < 0$
    c. $t > 0$
    d. $t < 0$
    e. $s_e \geq 0$

14. In a regression, it is always true that
    a. $b \geq 0$
    b. $b \leq 0$
    c. $s_b \geq 0$
    d. $s_b \leq 0$
    e. $r \leq 0$

15. In simple regression, s is equal to
    a. $s_e^2 / \Sigma(x_i - \bar{x})^2$
    b. s_e^2
    c. $s_e^2 / \Sigma(y_i - \bar{y})^2$
    d. $s_e^2 / [a / b]$
    e. $s_e^2[\Sigma x_i^2 / \Sigma(x_i - \bar{x})^2]$

16. In simple regression, *s* is equal to
    a. $s_e^2 / \Sigma(x_i - \bar{x})^2$
    b. s_e^2
    c. $s_e^2 / \Sigma(y_i - \bar{y})^2$
    d. $s_e^2 / [a / b]$
    e. $s_e^2[\Sigma x_i^2 / \Sigma(x_i - \bar{x})^2]$

17. For a regression consisting of n observations, an $1 - \alpha\%$ confidence interval for the slope coefficient b would be
    a. $b \pm t(\alpha, n - 2)\, s_b$
    b. $b \pm t(\alpha, n - 2) s_b^2$
    c. $b \pm t(\alpha/2, n - 2)\, s_b$
    d. $b \pm t(\alpha / 2, n - 2) s_b^2$
    e. $b \pm t(\alpha/2, n - 2)\, s_e$

18. For a regression consisting of n observations, an $1-\alpha\%$ confidence interval for the intercept term a would be
    a. $a \pm t(\alpha, n-2)\, s_a$
    b. $a \pm t(\alpha, n-2) s_a^2$
    c. $a \pm t(\alpha/2, n-2)\, s_a$
    d. $a \pm t(\alpha/2, n-2) s_a^2$
    e. $a \pm t(\alpha/2, n-2)\, s_e$

True/False (if false, explain why)

1. When we are testing the significance of the slope coefficient in a regression, the null hypothesis is usually $\beta = 1$.
2. A negative and significant slope coefficient implies an indirect relationship between the dependent and independent variables.
3. Given a level of significance of $\alpha = .05$, a two-tailed test of the significance of β will have a smaller absolute critical value than a one-tailed test.
4. A t-test can be used to test whether the correlation coefficient for two normally distributed random variables is significant.
5. Predictions or forecasts are one of the important uses of regression analysis.
6. The confidence interval for the mean response will get wider the farther we are from the mean of x.
7. The predicted value of y will always be equal to the actual value of y.
8. $t_b = b/s_b$
9. The t-statistic for a coefficient can never be negative.
10. In simple regression, a high R^2 would imply a significant slope coefficient.
11. In predicting the value of y using a regression, we are confident about the accuracy of the prediction if the place of prediction is far away from the center of our data.
12. The errors-in-variable problem is due to mis-measurement of the explanatory variable.
13. A negative t-statistic for the slope coefficient implies that the explanatory variable is not significant.
14. A t-test can be used to test the significance of the covariance between two random variables.

Questions and Problems

1. You are given the following information:

   $$\Sigma x_i^2 = 92, \quad \Sigma(x_i - \bar{x})^2 = 3.7, \quad s_e^2 = 1.25, \quad n = 40$$

   a. Compute s_a

   b. Compute s_b

2. Suppose you estimate the following regression of earnings against years of schooling using 30 observations. (Standard errors are reported in parentheses)

   $$\underset{}{EARN_i} = \underset{(4{,}825)}{11{,}929} + \underset{(127)}{421}\ SCHOOLING_i$$

   a. Test the significance of the slope coefficient at the 5% level of significance.

   b. Construct a 95% confidence interval for the slope coefficient.

3. Suppose in the regression in Question 2 the SSE = 75 and SSR = 81. Use an *F*-test to test the significance of the regression. Use a 5% level of significance.

4. Suppose you compute the correlation between two random variables to be .62. If 25 observations were used to estimate the correlation coefficient, use a 5% level of significance to test the significance of the correlation coefficient.

Answers to Supplementary Exercises

Multiple Choice

1. c	6. b	11. a	16. a
2. c	7. b	12. e	17. c
3. a	8. d	13. e	18. c
4. b	9. c	14. c	
5. c	10. b	15. e	

True/False

1. False. $\beta = 0$.
2. True
3. False. Larger absolute critical value.
4. True
5. True
6. True
7. False. The predicted value is just a guess, and therefore does not have to equal the actual value.
8. False. Unless H_0: $b_0 = 0$.
9. False. When the coefficient is negative, the *t*-value will be negative.
10. True
11. False. We become less confident as we move far away from the center of the data.
12. True
13. False. May be significantly negative.
14. False. Used to test the significance of the correlation coefficient.

Questions and Problems

1. a. $$s_a^2 = s_e^2 \frac{\Sigma x_i^2}{n\,\Sigma(x_i - \bar{x})^2} = 1.25\frac{92}{(40)(3.7)} = .777$$

$$s_a = \sqrt{.777} = .881$$

b. $$s_b^2 = \frac{s_e^2}{\Sigma(x_i - \bar{x})^2} = \frac{1.25}{3.7} = .338$$

$$s_b = \sqrt{.338} = .581$$

2. a. $t_b = 421/127 = 3.31$

$t_{.05/2,\,30-2} = 2.048$, so we reject H_0: $\beta = 0$.

b. $b \pm s_b t_{\alpha/2,\, n-2}$
$421 \pm 127(2.048)$

So the interval is from 160.90 to 681.10.

3. $$F_{1,n-2} = \frac{SSR/1}{SSE/(n-2)} = \frac{81/1}{75/(30-2)} = 30.24$$

$F_{.05,\,1,\,28} = 4.20$, so we reject H_0: $\beta = 0$.

4. $$t_{n-1} = \frac{r\sqrt{n-2}}{\sqrt{1-r^2}} = \frac{.62\sqrt{25-2}}{\sqrt{1-.62^2}} = 4.83$$

$t_{.05/2,\,24} = 2.064$, so reject H_0: $r = 0$.

CHAPTER 15

MULTIPLE LINEAR REGRESSION

Chapter Intuition

In previous chapters, we learned how regression analysis can be used to measure the relationship between two variables. In this chapter, we will learn how we can relate the dependent variable to two or more independent or explanatory variables. For example, suppose a farmer is interested in knowing what factors affect the amount of corn produced. In simple regression, he might assume that corn production depends on either rainfall or the amount of fertilizer. However, in real life, the farmer knows that both rainfall and the amount of fertilizer used determine the amount of corn.

In this case, he could use multiple regression to measure the relationship between a dependent variable (corn production) and two independent variables (rainfall and fertilizer). The slope coefficients for rainfall and fertilizer will measure how each factor affects corn production assuming no change in the other factor. That is, the coefficient for rainfall will measure the effect of changes in rainfall on corn production assuming we do not change the amount of fertilizer. Likewise, the coefficient for fertilizer will measure the effect of changes in the amount of fertilizer assuming no change in the amount of rainfall.

As in simple regression, our estimates of the coefficients are just educated guesses about the actual parameter value. To judge the significance of the coefficients we use statistical theory to test the significance of the estimates or to construct confidence intervals around the estimates.

One consideration that is not a concern in simple regression but may be a problem in multiple regression is ***multicollinearity***. Multicollinearity is a phenomenon where two explanatory variables are too highly related. When this problem occurs, it is impossible to separate their effects on the dependent variable. For example, serious multicollinearity may occur when we estimate a regression that explains an individual's wages based on experience and age. Your intuition tells you that there is a close relationship between age and experience, because as a person's age increases by one year, so does experience. In this case, if a person's wages increase by $500, it will be impossible to determine whether age or experience was the cause.

Chapter Review

1. ***Multiple linear regression*** is just a natural extension of the simple linear regression that was discussed in the previous chapters. In multiple regression, we assume that the dependent variable, y, can be influenced by more than one factor. For example, the amount of sales a company generates may depend on both the price it charges and the advertising dollars spent promoting the product.

2. The assumptions of the multiple linear regression model are:

   a. The error term is normally distributed with a mean of zero and a constant variance.

   b. The error terms are assumed to be independent of the k independent variables.

   c. Error terms are assumed to be independent of one another.

   d. The independent variables are not linearly related to each other. If they are, we say ***multicollinearity*** exists. Multicollinearity is sometimes a problem in multiple regression.

3. The same technique that was used to derive the simple linear regression model is also used in multiple linear regression, that is, we again minimize the squared error terms in order to find the "best" values for the slope coefficients. In simple regression, we fit a regression line to describe the relationship between x and y. In multiple regression, we fit a ***regression plane*** to describe this relationship.

4. The effect of each explanatory variable on the dependent variable is measured by the ***partial regression coefficient***. A partial regression coefficient measures the effect of the explanatory variable on the dependent variable, assuming that all other explanatory variables are held constant.

5. Because it is quite tedious to estimate the partial regression coefficients by hand, we usually use a computer program.

6. When testing the statistical significance of the slope coefficients, there are two ways we can do this. By using a t-test, we can test the significance of each individual slope coefficient. By using an F-test, we can jointly test the significance of all the slope coefficients simultaneously. Recall that the F and t-tests were equivalent in the simple linear regression model because there was only one slope coefficient. However, when we have more than one independent variable these tests are not equivalent.

Useful Formulas

Multiple regression model:

$$y_i = \alpha + \beta_1 x_{1i} + \beta_2 x_{2i} + \ldots + \beta_k x_{ki} + \varepsilon_i$$

Regression coefficients with two independent variables:

$$b_1 = \frac{\left(\sum_{i=1}^{n} x_{i2}^2 - n\,\overline{x}_2^2\right)\left[\sum_{i=1}^{n} x_{i1}\, y_i - n\overline{x}_1\,\overline{y}\right] - \left((\sum_{i=1}^{n} x_{i1}x_{i2} - n\overline{x}_1\,\overline{x}_2)\,(\sum_{i=1}^{n} x_{i2}\, y_i - n\overline{x}_2\overline{y})\right)}{\left(\sum_{i=1}^{n} x_{i1}^2 - n\overline{x}_1^2\right)\left(\sum_{i=1}^{n} x_{i1} - n\overline{x}_2^2\right)\left(\sum_{i=1}^{n} x_{i1}x_{i2} - n\overline{x}_1\overline{x}_2\right)^2}$$

$$b_2 = \left(\sum_{i=1}^{n} x_{i1}^2 - n\overline{x}_1^2\right)\left(\sum_{i=1}^{n} x_{i1} y_i - n\overline{x}_2\overline{y}\right) - \frac{\left((\sum_{i=1}^{n} x_{i1}x_{i2} - n\overline{x}_1\overline{x}_2)\,(\sum_{i=1}^{n} x_{i1} y_i - n\overline{x}_1\overline{y})\right)}{\left(\sum_{i=1}^{n} x_{i1}^2 - n\overline{x}_1^2\right)\left(\sum_{i=1}^{n} x_{i2}^2 - n\overline{x}_2^2\right)\left(\sum_{i=1}^{n} x_{i1}x_{i2}\; n\overline{x}_1\overline{x}_2\right)^2}$$

$$a = \overline{y} - b_1\overline{x}_1 - b_2\overline{x}_2$$

Coefficient of determination:

$$R^2 = 1 - \frac{SSE}{SST}$$

Adjusted coefficient of determination:

$$\overline{R}^2 = 1 - \frac{SSE/(n-k-1)}{SST/(n-1)} = 1 - (1 - R^2)\frac{(n-1)}{(n-k-1)}$$

F-ratio:

$$F_{k,n-k-1} = \frac{\sum_{i=1}^{n}(\hat{y}_i - \bar{y})^2}{\sum_{i=1}^{n}(y_i - \hat{y}_i)^2} \; \frac{n-k-1}{k}$$

Example Problems

Example 1 **Multiple Regression**

Suppose an NFL scout is interested in what physical attributes make for a good running back. He collects data on the weight and speed in the 40 yard dash of 7 running backs and their yards gained for the year. The data are summarized in the following table.

y Yards	x_1 Weight	x_2 Time in 40 yard dash
925	210	4.8
850	185	4.7
1622	225	4.7
1121	215	4.6
658	180	4.9
977	212	4.6
574	195	5.0

Estimate the regression coefficients, b_1 and b_2 and interpret the results.

Solution: For most multiple regression problems, it is easier to use a computer program to solve for the coefficients.

$\bar{y} = 961, \quad \bar{x}_1 = 203.14, \quad \bar{x}_2 = 4.66$

$\Sigma(x_{i1} - \bar{x}_1)^2 = 1{,}674.86 \qquad \Sigma(x_{i2} - \bar{x}_2)^2 = .4171$

$\Sigma(y_i - \bar{y})(x_{i1} - \bar{x}_1) = 28{,}417 \qquad \Sigma(y_i - \bar{y})(x_{i2} - \bar{x}_2) = -495.9$

$$\Sigma(x_{i1}-\bar{x}_1)(x_{i2}-\bar{x}_2)=-20.46$$

$$b_1 = \frac{(28{,}417)(.4171)-(-495.9)(-20.46)}{(1674.86)(.4171)-(-20.46)^2} = 6.101$$

$$b_2 = \frac{(-495.9)(1647.86)-(28{,}417)(-20.46)}{(1674.86)(.4171)-(-20.46)^2} = -889.602$$

$$a = \bar{y}-b_1x_1b_2x_2 = 961-6.101(203.143)-(-889.602)(4.66) = 3864.63$$

The coefficient b_1 indicates the effect of higher weight on the number of yards a running back gains. For every additional pound he weighs, he can expect to gain an additional 6 yards. The coefficient b_2 indicates the effect of increased speed on the number of yards a running back gains. For every 1/10 of second a running back can reduce his time in the 40 yard dash, he can expect to gain an additional 89 yards.

Example 2 **Significance of Regression Estimates**

Use the data and your results from Example 1 to test the significance of b_1 and b_2 at the 95% significance level.

Solution:

$$SSE = \Sigma(y_i-\hat{y}_i)^2 = 103{,}446.497$$

$$s_e^2 = SSE/(n-k-1) = 103{,}446.37\,/\,(7-2-1) = 25{,}861.59$$

$$s_{b1}^2 = \frac{s_e^2\ \Sigma(x_{i2}-\bar{x}_2)^2}{\Sigma(x_{i1}-\bar{x}_1)^2\ \Sigma(x_{i2}-\bar{x}_2)^2-[\Sigma(x_{i1}-\bar{x}_1)(x_{i2}-\bar{x}_2)]^2}$$

$$= \frac{25{,}861.59(.4171)}{1647.86(.4171)-\ 418.94^2} = 38.51$$

$$t_{b1} = \frac{b_1-0}{s_{b1}} = -\frac{6.101-0}{\sqrt{38.51}} = .9831$$

$$s_{b2}^2 = \frac{s_e^2 \sum(x_{i1}-\bar{x}_1)^2}{\sum(x_{i1}-\bar{x}_1)^2\sum(x_{i2}-\bar{x}_2)^2-[\sum(x_{i1}-\bar{x}_1)(x_{i2}-\bar{x}_2)]^2}$$

$$= \frac{25,861.59(1674.86)}{1647.86(.4171)-418.94^2} = 154,606.24$$

$$t_{b2} = \frac{b_2-0}{s_{b2}} = \frac{-889.602-0}{\sqrt{154,606.24}} = -2.26$$

Critical value for $t_{.025,4} = 2.776$, so neither variable is significant in explaining the number of yards a running back gains. However, b_2 is nearly significant, and would have been significant had we chosen a significance level of 10% instead of 5%.

Example 3 **Joint Significance Test**

Use the data and your results from Examples 1 and 2 to test the joint significance of b_1 and b_2 at a 5% level of significance.

Solution: A joint test means we are testing H_0: $\beta_1 = \beta_2 = 0$. The joint significance can be tested by using an *F*-test.

$F_{k,n-k-1} = [SSR/k]/[SSE/(n-k-1)]$

SST = SSE + SSR

$$SST = \sum(y_i - \bar{y})^2 = 717,972$$

$$SSE = \sum(y_i - \hat{y}_i)^2 = 103,446.497$$

$$SSR = 717,972 - 103,446.497 = 614,525.503$$

$F = [614,525.503/2]/[103,446.497/4] = 11.88$

The critical value for $F_{2,4} = 19.2$, so we are unable to reject the null hypothesis.

Example 4 **Interpreting Parameter Estimates**

Suppose a labor economist is interested in the relationship between experience and education on income. She estimates the following regression.

$$INCOME_i = 18{,}000 + 1{,}200\ EXPER_i + 800\ EDUC_i$$

where, $INCOME_i$ = income for person i measured in dollars

$EXPER_i$ = years of experience for person i

$EDUC_i$ = years of education for person i

Interpret the regression coefficients for EXPER and EDUC.

Solution: The regression says that a person is expected to earn $18,000 regardless of experience or amount of schooling. For each additional year of experience, we expect his/her income to rise by $1,200. For each additional year of schooling, we expect income to rise by $800.

Example 5 **Coefficient of Determination**

Suppose you estimate the following model

$$y_i = \alpha + \beta_1 x_{1i} + \beta_2 x_{2i} + e_i$$

based on 20 observations and obtain

$$\Sigma(y_i - \hat{y}_i)^2 = 285 \qquad \Sigma(y_i - \bar{y})^2 = 425$$

Compute the R^2. Does the model provide a good fit?

Solution:

$$SSE = \Sigma(y_i - \hat{y}_i)^2$$
$$SST = \Sigma(y_i - \bar{y})^2$$

$$R^2 = 1 - SSE/SST = 1 - 285/425 = .3294$$

An R^2 of .3294 means that the independent variables x_1 and x_2 explain 32.94% of the variation in the dependent variable y. Whether this is a good fit or not depends on what the dependent variable is. For example, being able to explain 32% of the variation in stock prices may be quite impressive, while being able to explain 32% of the variation in sales may not be very good.

Supplementary Exercises

Multiple Choice

1. In multiple regression there is
   a. more than one dependent variable but only one independent variable.
   b. more than one independent variable but only one dependent variable.
   c. more than one dependent variable and more than one independent variable.
   d. only one dependent variable and only one independent variable.
   e. more than two dependent variables and more than one independent variable.

2. When estimating a regression model with two explanatory variables, the geometric interpretation is that wc arc fitting
   a. a straight line to describe the relationship between the dependent variable and the explanatory variables.
   b. a triangle to describe the relationship between the dependent variable and the explanatory variables.
   c. a plane to describe the relationship between the dependent variable and the explanatory variables.
   d. a circle to describe the relationship between the dependent and explanatory variables.
   e. an ellipse to describe the relationship between the dependent and explanatory variables.

3. In multiple regression, each slope coefficient indicates
   a. the total influence of all the independent variables on the dependent variable.
   b. the influence of the independent variable on the dependent variable holding the other independent variables constant.
   c. where the regression plane intersects the y axis.
   d. both the partial and total influence of the independent variables.
   e. a point that is equal to the intercept term.

4. Multicollinearity occurs when
   a. the error term does not have a zero mean.
   b. the error term is not independent of the explanatory variables.
   c. two error terms are correlated with one another.
   d. two independent variables are highly correlated with one another.
   e. the variance of the error terms is not constant.

5. To test the significance of an individual slope coefficient we use
   a. an F-test.
   b. a t-test.
   c. a chi-square test.
   d. the binomial distribution.
   e. the exponential distribution.

6. To test the significance of an entire regression we use
   a. an F-test.
   b. a t-test.
   c. a chi-square test.
   d. the binomial distribution.
   e. the exponential distribution.

7. Other things being equal, as we increase the number of explanatory variables in a regression,
   a. R^2 increases.
   b. R^2 decreases.
   c. R^2 can increase or decrease.
   d. there is no effect on R^2.
   e. the probability of multicollinearity decreases.

8. The relationship between R^2 and adjusted R^2 is
   a. adjusted $R^2 = R^2$
   b. adjusted $R^2 = R^2(n-1)/(n-k-1)$
   c. adjusted $R^2 = [1-(1-R^2)](n-1)/(n-k-1)$
   d. adjusted $R^2 = 1-R^2$
   e. adjusted $R^2 = 1+R^2$

9. Adjusted R^2 is
   a. a measure of goodness of fit.
   b. adjusted for the use of additional explanatory variables.
   c. measures the percentage of the total variation of the dependent variable explained by the regression.
   d. generally a better measure of goodness of fit than R^2 in multiple regression.
   e. all of the above.

10. For a regression with n-observations and k-explanatory variables, the relationship between R^2 and F is
    a. $F_{k,\,n-k-1} = [(n-k-1)/k][R^2/(1-R^2)]$
    b. $F_{k,\,n-k-1} = R^2/(1-R^2)$
    c. $F_{k,\,n-k-1} = [n/k][R^2/(1-R^2)]$
    d. $F_{k,\,n-k-1} = [(n-k-1)/k]R^2$
    e. $F_{k,\,n-k-1} = R^2/(1+R^2)$

11. If you estimate an earnings equation for 100 individuals with the number of years of schooling as one explanatory variable and the number of months of school as a second explanatory variable you may
    a. get estimates that are not BLUE.
    b. suffer from serial correlation.
    c. have unequal variances of the error terms.
    d. suffer from multicollinearity.
    e. both b and d.

12. The degrees of freedom for the t-statistic used to test the significance of the slope coefficient in a regression consisting of 35 observations and three explanatory variables is
    a. 35
    b. 3
    c. 32
    d. 33
    e. 31

13. The degrees of freedom of the numerator in an F-test in a regression consisting of 50 observations and four explanatory variables is
    a. 50
    b. 4
    c. 3
    d. 46
    e. 45

14. The degrees of freedom of the denominator in an F-test in a regression consisting of 50 observations and four explanatory variables is
    a. 50
    b. 4
    c. 3
    d. 46
    e. 45

15. One problem that may occur in multiple regression that will not occur in simple regression is
    a. correlation between error terms.
    b. unequal variances of the error terms.
    c. correlation between the error terms and the explanatory variables.
    d. correlation between explanatory variables.
    e. an error term without a zero mean.

True/False (if false, explain why)

1. When multicollinearity is present, the least squares estimator is still BLUE.
2. It is possible to have insignificant t-values for the slope coefficients and a significant F-value for the regression.
3. If two explanatory variables are perfectly correlated, it will be impossible to use the least squares method.
4. R^2 is a better measure of the fit of a regression than adjusted R^2 in multiple regression.
5. If the correlation between each pair of explanatory variables is below .5, multicollinearity will never be a problem.
6. A t-statistic can never be negative.

7. A negative t-statistic on one of the slope coefficients implies that this regressor is not important.
8. Simple regression is just a special case of multiple regression.
9. The geometric interpretation of a regression consisting of two explanatory variables is a straight line.
10. Multicollinearity occurs when the explanatory variables are highly correlated with the dependent variable.
11. Adjusted R^2 is always larger than R^2.
12. A t-test is used to test the joint significance of the slope coefficients in a multiple regression.
13. An F-test is used to test the significance of individual slope coefficients.
14. SST = SSR + SSE.
15. SST can never be negative.

Questions and Problems

1. Suppose you are an economist with the Department of Labor and are interested in examining the relationship between earnings and other factors such as age, education level and work experience.
   a. Which of the above variables should be the dependent variable? Which variables could serve as explanatory variables?
   b. Are there any possible problems you might encounter in estimating your regression?
2. Suppose you estimate a regression of stock returns for a company against the earnings per share (EPS) and debt/equity ratio for the company.
   a. Write down the regression model that should be estimated.
   b. Interpret the meaning of the slope coefficients.
3. Suppose you estimate a regression using three explanatory variables that have the following relationship:

$$x_{3i} = 2x_{1i} + x_{2i}$$

   a. Are you likely to encounter any problems in estimating a regression using all three explanatory variables?

b. Is there any way to solve this problem?

4. Suppose you estimate a regression using 50 observations and four explanatory variables. The R^2 for the regression is .64. Compute the adjusted R^2 for the regression.

5. Use an F-test to jointly test the significance of the slope coefficients for the regression given in Problem 4 at the 5% level of significance.

6. You estimate the following regression using 30 observations.

$$y_i = 122 + 32.4x_{1i} - .78x_{2i}$$
$$r_{x1,x2} = .59$$
$$\Sigma(x_{i1} - \bar{x}_1)^2 = 23.6$$
$$\Sigma(x_{i2} - \bar{x}_2)^2 = 12.3$$

$$s = 41.7$$

Test the significance of the slope coefficients at the 5% level.

7. Use the information and your calculations from Problem 6 to construct 95% confidence intervals for the slope coefficients.

Answers to Supplementary Exercises

Multiple Choice

1. b	6. a	11.d
2. c	7. a	12.e
3. b	8. c	13.b
4. d	9. e	14.e
5. b	10. a	15.d

True/False

1. True
2. True
3. True

4. False. Adjusted R^2 is better in multiple regression because it adjusts for the number of explanatory variables.
5. False. It's possible for two or more explanatory variables to form a linear combination of another explanatory variable, thus leading to multicollinearity.
6. False. The t-value will be negative when the coefficient is negative.
7. False. Can be significantly negative.
8. True
9. False. The geometric interpretation is a plane.
10. False. Occurs when the explanatory variables are high correlated with one another.
11. False. Adjusted R^2 is always less than or equal to R^2.
12. False. An F-test is used to test joint significance.
13. False. A t-test is used to test individual coefficients.
14. True
15. True

Questions and Problems

1. a. Earnings would be the dependent variable, and age, work experience, and education level would be the explanatory variables.

   b. Because a person's age and work experience are likely to be highly correlated, multicollinearity may be a problem.

2. a. $RETURN_t = \alpha + \beta_1 EPS_t + \beta_2 DE_t$

   b. The slope coefficient, β_1, measures the impact of changes in earnings per share on the stock return, assuming that the debt/equity ratio remains constant. The slope coefficient, β_2, measures the impact of changes in the debt/equity ratio on the stock return, assuming that the EPS remains constant.

3. a. Because x_3 is a linear combination of x_1 and x_2, multicollinearity will be a problem.

   b. One method for dealing with this problem is to drop one of the explanatory variables from the regression.

4. Adjusted $R^2 = 1 - (1 - R^2)[(n - 1)/(n - k - 1)]$
   $= 1 - (1 - .64)[(50 - 1)/(50 - 4 - 1)] = .608$

5. $F_{k, n-k-1} = [(n - k - 1)/k][R^2/(1 - R^2)]$
   $= [(50 - 4 - 1)/4][.64/(1 - .64)] = 20$

   $F_{.05, 4, 45}$ is approximately 2.4, so we can reject the null hypothesis that all slope coefficients are jointly equal to 0.

6. $$s_{b1}^2 = \frac{s_e^2}{(1-r_{x1,x2}^2)\Sigma(x_{i1}-\bar{x}_1)^2} = \frac{41.7}{(1-.59)(23.6)} = 4.31$$

$$s_{b1} = \sqrt{4.31} = 2.08$$

$$s_{b2}^2 = \frac{s_e^2}{(1-r_{x1,x2}^2)\Sigma(x_{i2}-\bar{x}_2)^2} = \frac{41.7}{(1-.59)(12.3)} = 8.27$$

$$s_{b2} = \sqrt{8.27} = 2.88$$

$t_{b1} = 32.4/2.08 = 15.58$
$t_{b2} = -.78/2.88 = -2.71$

$t_{.05,\,27} = 2.052$, so reject the null hypothesis of $\beta_1 = 0$, but accept the null hypothesis of $\beta_2 = 0$.

7.

$b_1 \pm s_{b1}\ t_{\alpha/2,\,n\text{-}k\text{-}1}$	$b_2 \pm s_{b2}\ t_{\alpha/2,\,n\text{-}k\text{-}1}$
$32.4 \pm 2.08(2.052)$	$-.78 \pm 2.88(2.052)$
28.13 to 36.67	−6.69 to 5.13

CHAPTER 16

OTHER TOPICS IN APPLIED REGRESSION ANALYSIS

Chapter Intuition

In the preceding chapters, we learned about linear regression. One of the things which we learned is that for linear regression to be valid, certain rules or assumptions must hold. What problems will occur when the rules are violated and how can we deal with these problems is the focus of this chapter.

Previously, we introduced the concept of ***multicollinearity***. Multicollinearity occurs when explanatory variables are closely related. When this occurs, it will be impossible to separate their impact on the dependent variable.

A second problem that can occur is ***heteroscedasticity***. Heteroscedasticity occurs when the variance of the error terms is not constant. For example if we were estimating a regression that explains an individual's consumption based on his or her income, it would be reasonable to assume that the error terms will not have a constant variance because wealthier people will have a higher variance in their consumption. Heteroscedasticity, if it occurs, causes the least squares method to give more weight to high variance observations, and thus causes the model to have a greater variance than it otherwise would.

Autocorrelation occurs when error terms are correlated with previous error terms. This problem will lead to a regression model that is not efficient.

This chapter also tells us how to deal with variables which are qualitative in nature. For example, if an economist believes that one factor influencing a person's earnings is the sex of the individual, she can use a dummy variable to capture this difference.

Chapter Review

1. Previously, you learned that the simple and multiple linear regression models were based on several assumptions. When these assumptions are violated, there may be problems in the estimation of the coefficients or in statistical inference.

2. ***Multicollinearity*** results when two or more independent variables are highly correlated with another, or when one of the independent variables can be written as a linear combination of some of the remaining independent variables.

   a. If the independent variables are perfectly correlated, the least squares approach cannot be used to estimate the regression coefficients.

   b. If the independent variables are highly but not perfectly correlated, there will be an increase in the standard error of the coefficients and hence the t-values will be extremely small.

   c. Detecting multicollinearity can be difficult. However, two simple methods may enable us to detect multicollinearity. First, we can examine the correlation between each pair of independent variables. If any pair has a correlation of .8 or .9, multicollinearity may be a problem. The second method is to look at the t-values for the regression coefficients and the F-value for the entire regression. If the t-values are insignificant while the F-value is significant, a problem of multicollinearity may exist.

3. ***Heteroscedasticity*** occurs when the variance of the error terms in the regression model is not constant. When heteroscedasticity is present, the parameter estimates will still be unbiased; however, the parameter estimates will be inefficient. Heteroscedasticity is a common problem in cross-sectional regressions.

   a. One way to detect heteroscedasticity is to look at a plot of the error terms against the independent variables. If a constant relationship does not appear to hold, heteroscedasticity may be a problem. In the following figure, we can see that the variance or dispersion of the error terms gets larger as the size of the independent variable increases.

   b. A second method for detecting heteroscedasticity is to run a regression using the squared error terms as the dependent variable against the independent variables. If any of the t-values are significant, heteroscedasticity exists.

   c. When heteroscedasticity is a problem, we sometimes estimate the model using a method known as weighted least squares.

4. ***Autocorrelation*** occurs when the error terms are correlated with past values of the error term. When autocorrelation is present, the parameter estimates will be unbiased but inefficient. Positive autocorrelation occurs when high values for the error terms tend to be followed by high values, and when low values for the error terms tend to be followed by low values as in the following figure.

   Negative autocorrelation occurs when high values for the error terms tend to be followed by low values, and vice versa as in the following figure.

a. Autocorrelation can be detected by using the ***Durbin-Watson statistic*** (DW) and checking the DW table at the end of the text (Table A9). The problem with using the DW statistic is that the test results may be inconclusive depending on the value of the computed DW statistic. To use the Durbin-Watson statistic we use the values for the upper DW value d_U and the lower DW value d_L and follow the following rules:

   i. For a one-tailed test of H_0: no autocorrelation vs. H_1: positive autocorrelation. We will reject H_0 if $d < d_L$. We will accept H_0 if $d > d_U$. The test will be inconclusive if $d_L \leq d \leq d_U$.

   ii. For a one-tailed test of H_0: no autocorrelation vs. H_1: negative autocorrelation. We will reject H_0 if $d > 4 - d_L$. We will accept H_0 if $d < 4 - d_U$. The test will be inconclusive if $4 - d_U \leq d \leq 4 - d_U$.

   iii. For a two-tailed test of H_0: no autocorrelation vs. H_1: positive or negative autocorrelation. We will reject H_0 if $d < d_L$ or $d > 4 - d_L$. We will accept H_0 if $d_U < d < 4 - d_U$. The test will be inconclusive if $d_L \leq d \leq d_U$ or $4 - d_U \leq d \leq 4 - d_U$.

b. When a lagged dependent variable is included in the regression model, the DW statistic will not be valid. In this case, it is necessary to use a different statistic known as ***Durbin's H***.

5. ***Specification error*** results when the regression model is incorrectly specified. This can result from the omission of a relevant variable or the inclusion of an irrelevant variable.

6. Sometimes there is not a linear relationship between the x and y variables. In this case, a nonlinear model may be a better choice for the regression. The simplest nonlinear model is the quadratic model in which a squared value of the independent variable is included as an independent variable in order to capture the nonlinear relationship.

7. When we estimate a regression over time, we sometimes believe that the current value of y, y_t depends on past values of y such as y_{t-1} or y_{t-2}. In this case, we can use a lagged dependent variable as an independent variable in order to capture this effect.

8. In many instances, we are interested in incorporating some qualitative information into our regression, like the sex of the worker or geographic region of the country. When this is the case we can place a binary variable known as a ***dummy variable*** into the regression so we can capture these effects.

9. When two independent variables are assumed to work together in determining the value of the dependent variable as in the case of rainfall and fertilizer on crop production, we can capture this effect by using an interaction variable. An interaction variable can be created by multiplying the two independent variables together and using this new variable as an additional independent variable in the regression.

Useful Formulas

Durbin-Watson statistic:

$$DW = \frac{\sum_{t=2}^{n}(e_t - e_{t-1})^2}{\sum_{t=1}^{n} e_t^2}$$

Durbin's H:

$$H = \left[1 - \frac{d}{2}\right]\sqrt{\frac{n}{1 - n\,V(b_1)}}$$

Quadratic regression model:

$$y_i = \alpha + b_1X_1 + b_2X_1^2 + e_i$$

Lagged dependent variable model:

$$y_t = \alpha + \beta_1X_{1t} + \beta_2X_{2t} + \ldots + \beta_kX_{kt} + \gamma y_{t-1} + e_t$$

Dummy variable model:

$$y_i = \alpha + \beta_1X_{1i} + \beta_2X_{2i} + \ldots + \beta_kX_{ki} + \gamma D_{1i} + e_i$$

Interaction variable model:

$$y_t = \alpha + \beta_1X_{1t} + \beta_2X_{2t} + \beta_3(X_{1t} \times X_{2t}) + e_t$$

Example Problems

Example 1 **Durbin-Watson Statistic**

Suppose you have a sample of 25 observations and two explanatory variables, and you want to test for autocorrelation. What can you say about autocorrelation for each of the following Durbin-Watson statistics?

a. d = 1.20
b. d = 2.00
c. d = 3.25
d. d = 1.55
e. d = 2.55

Solution: In order to determine whether or not autocorrelation is a problem, we need to examine the table of Durbin-Watson values. Remember, the DW statistic has upper and lower values, and a range where the test is inconclusive. For 25 observations, 2 explanatory variables and a .05 significance level, $d_L = 1.21$, $d_U = 1.54$.

a. negative autocorrelation because $d < 1.21$.

b. no autocorrelation because $1.21 < d < 4 - 1.54$.

c. negative autocorrelation because $d > 4 - 1.21$.

d. no autocorrelation because $1.21 < d < 4 - 1.54$.

e. inconclusive because $4 - 1.54 \leq d \leq 4 - 1.21$.

Example 2 **Multicollinearity**

You are interested in the relationship between y and three possible explanatory variables x_1, x_2 and x_3. You are given the following correlation matrix

	y	x_1	x_2	x_3
y	1.00	.55	.66	.89
x_1		1.00	.82	.75
x_2			1.00	.65
x_3				1.00

Given the above information, does multicollinearity appear to be a problem? If so, between which variables?

Solution: In the table, we have the correlation between each pair of variables. Because x_1 and x_2 have a correlation of .82, multicollinearity may be a problem.

Example 3 **Dummy Variables**

Suppose you have been hired by a lawyer who is interested in showing that a company discriminates against women in the wages they pay. You estimate the following regression:

$$WAGEi = 19{,}000 + \underset{(821)}{2{,}000}\ EXPERi + \underset{(332)}{1{,}000}\ EDUCi + \underset{(3{,}400)}{4{,}000}\ SEXi$$

where WAGEi = wage for person i
EXPERi = years of experience for person i
EDUCi = years of education
SEXi = dummy variable = 1 for female
= 0 for male

Standard errors of the coefficients are reported in the parentheses.

a. Interpret the coefficients for experience and education.

b. Interpret the coefficient for sex. Does discrimination exist?

Solution: a. The coefficients for experience and education are both positive and significant (their t-ratios will exceed 2.00), showing workers with more experience and/or education get higher wages.

b. When we are looking for discrimination we are looking for a significant coefficient on our dummy variable for sex. A positive and significant coefficient would imply that females earn more than males with similar education and experience. A negative coefficient would imply that females earn less than males with similar education and experience. Because the coefficient on sex is not significant, we cannot conclude that discrimination exists.

Example 4 **Computing Durbin-Watson Statistic**

You are given the following error terms from a regression

Period	e	Period	e
1	21.2	9	–9.6
2	– 15.3	10	10.2
3	12.4	11	–13.3
4	–21.0	12	7.4
5	9.4	13	–15.1
6	–25.5	14	12.0
7	– 8.4	15	–6.0
8	12.3	16	8.0

Compute the Durbin-Watson *d* statistic. Does autocorrelation appear to be a problem?

Solution:

$$d = \Sigma_{t=2}(e_t - e_{t-1})^2 / \Sigma e_t^2$$

$$\begin{aligned}\Sigma_{t=2}(e_t - e_{t-1})^2 &= (-15.3 - 21.2)^2 + (12.4 - (-15.3))^2 + \\ &\quad (-21 - 12.4)^2 + \ldots + (8 - (-6))^2 \\ &= 9{,}691.22\end{aligned}$$

$$e_t^2 = 21.2^2 + (-15.3)^2 + \ldots + 8^2 = 2{,}689.17$$

$$d = 9{,}691.22/2{,}689.17 = 3.60$$

From the table we know for $n = 16$ and $\alpha = .05$, $d_L = 1.10$, and $d_U = 1.37$. So negative autocorrelation exists because $d > 4 - d_L$.

Example 5 **Interaction Variables**

A biologist is interested in the effect of temperature and humidity on cell growth. She collects the following information from 6 samples.

Sample	Temperature	Humidity	Cells
1	4°	8%	10,000
2	8	12	11,122
3	12	19	14,025
4	16	30	19,022
5	20	50	26,872
6	24	75	42,308

Estimate the relationship between cells and temperature and humidity. Use an interaction variable to estimate the interaction effect of temperature and humidity on cell growth. Interpret your results.

Solution: To capture the interaction effect of temperature and humidity on cell growth, we create a third independent variable by multiplying temperature and humidity. The coefficient on this variable will capture this interaction effect. The regression we are interested in estimating is

$$Cell_i = a + b_1\, Temp_i + b_2\, Humid_i + b_3\, (Temp_i \times Humid_i)$$

Once we have created the interaction variable, we estimate the regression using the multiple regression procedure previously discussed. Because there are three independent variables we used the computer to estimate the regression. The results are:

$$Cell_i = 10{,}916.78 + \underset{(268.86)}{447.48}\, Temp_i - \underset{(321.04)}{502.65}\, Humid_i + \underset{(9.53)}{32.39}\, (T_i \times H_i)$$

$R^2 = .999$

$s_e = 540.9$

Standard errors are reported in parentheses.

Example 6 **Heteroscedasticity**

A financial analyst is interested in the relationship between dividend per share (DPS) and earnings per share (EPS). He collects information on these two variables. He estimates the following regression:

$$DPS_i = \alpha + \beta\, EPS_i + e_i$$

From this regression he computes the error from the regression for each company.

Company	EPS	e	Company	EPS	e
1	$.80	–.05	7	2.20	–2.12
2	1.10	.35	8	2.40	2.50
3	1.20	–.10	9	2.75	–3.00
4	1.30	.72	10	3.00	2.80
5	1.45	–.25	11	3.10	–3.50
6	2.05	1.21	12	3.40	4.00

Does heteroscedasticity appear to be a problem in this regression?

Solution:

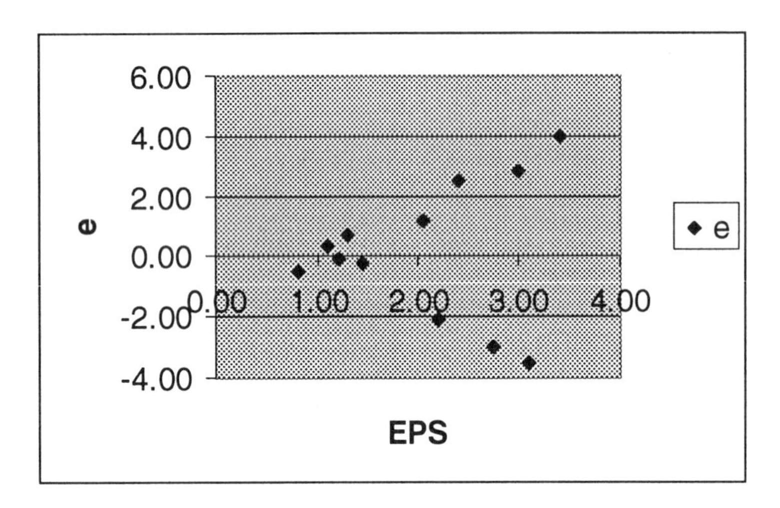

Because the error terms become larger as the independent variable increases in size, we have a heteroscedasticity problem.

Supplementary Exercises

Multiple Choice

1. Multicollinearity occurs when
   a. two or more independent variables are highly correlated.
   b. the variance of the error terms is not constant.
   c. current and lagged error terms are correlated.
   d. the independent variable is measured with error.
   e. we estimate an incorrect version of the true model.

2. Heteroscedasticity occurs when
   a. two or more independent variables are highly correlated.
   b. the variance of the error terms is not constant.
   c. current and lagged error terms are correlated.
   d. the independent variable is measured with error.
   e. we estimate an incorrect version of the true model.

3. Autocorrelation occurs when
   a. two or more independent variables are highly correlated.
   b. the variance of the error terms is not constant.
   c. current and lagged error terms are correlated.
   d. the independent variable is measured with error.
   e. we estimate an incorrect version of the true model.

4. Specification error occurs when
   a. the dependent variable is measured with error.
   b. we estimate an incorrect version of the true model.
   c. two or more independent variables are highly correlated.
   d. the variance of the error terms is not constant.
   e. two or more dependent variables are highly correlated.

5. Heteroscedasticity gives us
   a. biased parameter estimates.
   b. best linear unbiased estimates (BLUE).
   c. efficient parameter estimates.
   d. problems in statistical inference.
   e. a high degree of correlation between the error terms and the dependent variable.

6. Autocorrelation gives us
   a. biased parameter estimates.
   b. best linear unbiased estimates (BLUE).
   c. inefficient parameter estimates.
   d. a short cut to statistical inference.
   e. a high degree of correlation between the error terms and the independent variable.

7. Multicollinearity gives us
   a. biased parameter estimates.
   b. best linear unbiased estimates (BLUE).
   c. inefficient parameter estimates.
   d. problems in statistical inference.
   e. two error terms that are correlated with each other.

8. Specification error gives us
   a. biased parameter estimates.
   b. best linear unbiased estimates (BLUE).
   c. inefficient parameter estimates.
   d. multicollinearity.
   e. autocorrelation.

9. A dummy variable can be used when
   a. the independent variable is quantitative.
   b. the independent variable is qualitative.
   c. the independent variable is heteroscedastic.
   d. there is autocorrelation.
   e. multicollinearity is known not to be a problem.

10. If we are interested in using dummy variables to capture the effect of different months on stock returns we should use
    a. 12 dummy variables, one for each month of the year.
    b. 1 dummy variable.
    c. 11 dummy variables.
    d. as many dummy variables as we like.
    e. we cannot use dummy variables to capture this effect.

11. If we are estimating a U-shaped relationship such as the total cost curve in economics, we can best capture this relationship by using
    a. dummy variables.
    b. simple regression.
    c. a lagged dependent variable.
    d. a quadratic regression model.
    e. this relationship cannot be modelled using regression analysis.

12. If we use a log-log linear model to estimate the demand for ice cream, the slope coefficients would be
    a. elasticities.
    b. identical to the results we would get from a non-log model.
    c. always positive.
    d. an example of multicollinearity.
    e. impossible to interpret.

13. To correct the problem of heteroscedasticity, we can
    a. omit one or more of the independent variables.
    b. use a log transformation.
    c. use dummy variables.
    d. use weighted least squares.
    e. first correct for autocorrelation.

14. To correct the problem of multicollinearity, we can
    a. use a log transformation.
    b. omit one or more of the independent variables.
    c. use dummy variables.
    d. use weighted least squares.
    e. use a lagged dependent variable.

15. The use of a dummy variable to measure differences in earnings between males and females assumes that
    a. the slope coefficient for males and females will be the same.
    b. the intercept terms for males and females will be the same.
    c. both the slope coefficient and the intercept terms will be the same.
    d. both the slope coefficient and the intercept terms will be different.
    e. multicollinearity is a problem.

True/False (if false, explain why)

1. Heteroscedasticity leads to biased parameter estimates.
2. Autocorrelation leads to biased parameter estimates.
3. Heteroscedasticity can be detected using the Durbin-Watson statistic.
4. When perfect multicollinearity exists, it will be impossible to use the least squares method.
5. The Durbin-Watson statistic can be used to detect autocorrelation when a lagged dependent variable is included in regression.
6. Dummy variables can be used when qualitative variables such as sex or race are included in a regression.
7. Multicollinearity may be a problem when an independent variable is highly correlated with the dependent variable.
8. Omitting a relevant explanatory variable in a regression leads to specification error.
9. Including an irrelevant explanatory variable in a regression leads to specification error.
10. Using explanatory variables that are measured with error leads to specification error.
11. Including an irrelevant explanatory variable in a regression is worse than omitting a relevant variable.

12. The problem of heteroscedasticity can sometimes be detected by examining the correlation between explanatory variables.
13. The Durbin-Watson test will always tell us whether autocorrelation is a problem.
14. An interaction variable can be useful when we believe that the effect of an independent variable on the dependent variable will depend on the value of another independent variable.
15. One way to detect heteroscedasticity is to look at a graph of the residuals against the independent variables or the expected values.

Questions and Problems

1. Find d_U and d_L for a regression that has k explanatory variables and n observations:

   a. $k = 1, \ n = 25$

   b. $k = 2, \ n = 30$

   c. $k = 3, \ n = 20$

2. Suppose you run a regression using three explanatory variables and 40 observations. If you compute the Durbin-Watson statistic to be 1.25, is autocorrelation a problem?

3. Suppose you estimate a regression using one explanatory variable and 50 observations and compute the following:

$$\sum_{t=2}^{50} (\hat{\varepsilon}_t - \hat{\varepsilon}_{t-1})^2 = 6.27, \quad \sum_{t=1}^{50} \hat{\varepsilon}_t^2 = 2.71$$

   a. Compute the Durbin-Watson statistic.

   b. Is autocorrelation a problem?

4. Below is the correlation for the variables in a regression with two explanatory variables.

	y	x_1	x_2
y	1.00	0.98	0.75
x_1		1.00	0.85
x_2			1.00

Does multicollinearity appear to be a problem?

5. What possible problems might you encounter in the following regressions?

   a. A regression of IBM's stock returns against its earnings per share (EPS) for the last 10 years.

   b. A regression of the EPS for 100 different companies against the size of each company in 1993.

   c. A regression of the consumption of 5,000 people against each person's income in 1993.

6. Suppose you estimate an earnings equation based on experience, education and the sex of the individual. The equation you estimate is

$$EARN_i = 12{,}212 + 722\ EXPER_i + 927\ EDUC_i - 2{,}211\ SEX_i$$

where $SEX_i = 1$ for females and 0 for males.

Interpret the coefficient on the dummy variable for sex, SEX_i.

Answers to Supplementary Exercises

Multiple Choice

1. a	6. c	11. d
2. b	7. c	12. a
3. c	8. a	13. d
4. b	9. b	14. a
5. d	10 c	15. a

True/False

1. False. Heteroscedasticity leads to inefficient estimates.
2. False. Autocorrelation leads to inefficient estimates.
3. False. Autocorrelation is detected using the Durbin-Watson test.
4. True
5. False. Durbin's H should be used when a lagged dependent variable is used in a regression.
6. True
7. False. Multicollinearity occurs when independent variables are highly correlated.
8. True
9. False. A t-test may detect the irrelevance of a variable.
10. True
11. False. It is worse to omit a relevant explanatory variable than to include an irrelevant one.
12. False. Heteroscedasticity can be detected by examining a plot of the error terms from the regression.
13. False. The Durbin-Watson test has a range where the test is indeterminate.
14. True
15. True

Questions and Problems

1. a. $d_L = 1.29$, $d_U = 1.45$ c. $d_L = 1.00$, $d_U = 1.68$

   b. $d_L = 1.28$, $d_U = 1.57$

2. $d_L = 1.25$, $d_U = 1.68$

   $0 < d < d_L$, so we reject the null hypothesis of no autocorrelation and accept the alternative hypothesis of positive autocorrelation.

3. a. $d = 6.27/2.71 = 2.31$

   b. $d_L = 1.50$, $d_U = 1.59$

   $2 < d < 4 - d_U$, so we accept the null hypothesis of no autocorrelation.

4. The correlation between x_1 and x_2 is 0.85, so multicollinearity may be a problem.

5. a. In a time series regression like this, autocorrelation may be a problem.

   b. In a cross-sectional regression like this, heteroscedasticity may be a problem because we would expect larger companies to have higher variance in their EPS.

   c. Heteroscedasticity is likely to be a problem because individuals with higher incomes are likely to have a greater variance in their consumption patterns.

6. The coefficient on the dummy variable measures the difference in earnings between males and females, assuming a similar relationship between the experience and education of the individual and his or her earnings. Because the coefficient is –2,211, females are expected to earn $2,211 less than males with similar levels of education and experience.

CHAPTER 17

NONPARAMETRIC STATISTICS

Chapter Intuition

So far, you have learned how to conduct statistical analysis under conditions when the distribution of the sample is known. When we know the distribution of the sample we are using ***parametric statistics***. However, in many instances, we will not know the distribution of the sample. In this case, we need to take a different approach in order to conduct our analysis. This approach is known as ***nonparametric statistics*** because we do not know the parameters of the distribution. For example, we learned about ANOVA, which tests whether three or more means equal each other. An important assumption for conducting the test is that the data are drawn from a normal distribution. In this chapter, we will learn about a test known as the Kruskal-Wallis test which is used for conducting ANOVA without assuming that the data are drawn from a normal distribution. Many of the nonparametric tests discussed in this chapter use the "ranking" of data to make comparisons rather than comparing the data directly.

Chapter Review

1. ***Nonparametric tests*** are distribution free tests, which means that the distribution of the test statistic does not depend on the assumption that the data are drawn from a certain distribution.
2. The ***sign test*** compares the averages of two populations. It can also be used to examine the central tendency (median) of a population. The sign test uses the sign of the difference between pairs of numbers coming from different groups of data. In conducting a sign test, we utilize the test of a proportion. The statistic has a standard normal distribution.
3. The ***Wilcoxon matched-pairs signed-rank test*** compares the average of two populations. The data used in this test have to be matched pairs. This test uses not only the sign of the difference between a pair of numbers but also the quantitative

measurements of the difference. The test statistic follows a standard normal distribution when the sample is large.

4. The ***Mann-Whitney U test*** compares the average of two populations using the rank sum of the sample groups. Using this test does not requires matched pair data as in the Wilcoxon matched-pairs signed-rank test. The Mann-Whitney U test follows a standard normal distribution when the sample is large.

5. ***Spearman's rank correlation*** is used to measure whether the rankings of two variables are correlated. This statistic will generate a number between –1 and 1. A negative Spearman's rank correlation indicates that the rankings of the two variables move in different directions. That is, a higher ranking in one variable implies a lower ranking in the other variable. A *t*-test can be used to test the population rank correlation. The degrees of freedom are the sample size minus two.

6. The ***Kruskal-Wallis test*** is used to examine whether three or more averages are equal. The Kruskal-Wallis test computes the rank sum of each data group. The statistic follows a chi-square distribution when the sample is large. The degrees of freedom is the number of data groups minus one.

7. A ***runs test*** is used to study the randomness of data. In conducting the test, we need to count the number of runs, which is defined as consecutive numbers with the same sign. The runs test follows a standard normal distribution.

Useful Formulas

Mann-Whitney U test:

$$U = n_1 n_2 + \frac{n_1(n_2 + 1)}{2} - R_1$$

$$U = n_1 n_2 + \frac{n_2(n_2 + 1)}{2} - R_2$$

Kruskal-Wallis test:

$$K = \left[\frac{12}{n(n + 1)}\right] \sum_{i=1}^{k} \left[\frac{R_i^2}{n_i}\right] - 3(n+1)$$

Spearman's rank correlation:

$$r_s = 1 - \frac{6 \sum_{i=1}^{n} d_i^2}{n(n^2 - 1)}$$

Runs test:

$$\mu_R = \frac{2n_1 n_2}{n_1 n_2} + 1$$

$$\sigma_R^2 = \frac{2n_1 n_2 (2n_1 n_2 - n)}{n^2 (n-1)}$$

$$Z = \frac{R - \mu_R}{\sigma_R}$$

Example Problems

Example 1 **Testing the Equality of Two Means**

A statistics professor is interested in seeing if there is any difference between the average scores of those students who own the workbook and those who don't. The rankings are summarized below (a higher ranking means a higher grade):

Own			Don't Own		
2	4	6	1	3	5
7	8	15	9	10	11
18	17	16	14	13	12
19	20	26	21	22	23
30	28	27	29	25	24

Do a test to determine if students who own the workbook have higher average scores than students who do not own the workbook. Use a 5% level of significance.

Solution: H_0: $\mu_1 - \mu_2 \leq 0$
H_1: $\mu_1 - \mu_2 > 0$

where μ_1 = the average for students with the workbook.
μ_2 = the average for students without the workbook.

$$\frac{w_1 - \frac{n_1(n_1 + n_2 + 1)}{2}}{\sqrt{\frac{n_1 n_2 (n_1 + n_2 + 1)}{12}}} = \frac{243 - \frac{15(15+15+1)}{2}}{\sqrt{\frac{15(15)(15+15+1)}{12}}} = 0.44 < 1.65$$

We do not have enough evidence to reject the null hypothesis that students owning the workbook have a higher average.

Example 2 **Testing the Equality of Proportions**

A marketing company is conducting a survey of cola popularity at the mall. Forty people have been asked to taste two different brands of cola, Cola A and Cola B. These people are then asked to score each cola on a scale of 1 to 4, with 4 being the best flavor. The table below presents how each person scored the two colas.

A	B	A	B	A	B	A	B
1	2	3	4	1	3	4	1
2	1	3	2	1	2	4	2
4	3	1	2	3	2	4	2
1	3	4	1	2	3	2	4
2	4	4	3	4	2	4	1
2	4	2	1	4	2	3	2
2	4	3	4	4	2	2	1
4	3	4	2	1	2	2	3
4	2	3	1	4	2	3	1
4	3	2	1	2	4	4	2

Test whether the two brands of cola are equally popular. Use a 10% level of significance.

Solution: H_0: equally popular (p = 1/2)
H_1: not equally popular (p ≠ 1/2)

where p = the proportion of people who like Cola A better.

$$\frac{\bar{p}-p}{\sqrt{\frac{p(1-p)}{n}}} = \frac{25/40-1/2}{\sqrt{\frac{1/2(1-1/2)}{40}}} = 1.58 < 1.645 = z_{.10/2}$$

We do not have enough evidence to reject the null hypothesis.

Example 3 **Wilcoxon Matched-Pair Signed-Rank Test**

The manager of a fast food chain is interested in the effect of coupons on sales. He collected sales data from ten fast food restaurants under his supervision. The sales numbers are for June, when the coupons are in effect, and for July when there are no coupons.

(1) Restaurant	(2) June	(3) July	d (3)–(2)	rank of |d|	(+)	(–)
1	145	132	–13	7		7
2	156	160	4	3.5	3.5	
3	143	131	–12	8		8
4	155	132	–23	9		9
5	162	160	–2	1		1
6	159	159	0			
7	160	165	5	5	5	
8	149	152	3	2	2	
9	162	153	–9	6		6
10	144	140	–4	3.5		3.5
Total					10.5	34.5

Do the data support the hypothesis that coupons make a difference? Use a 5% level of significance.

Solution: $w^+ = 10.5$, $w^- = 34.5$. There are nine differences in our data. The two-tailed value at $n = 9$ and $\alpha = .05$ is 6. Because the value of w^+ is higher than 6, we reject the null hypothesis of no difference.

Example 4 **Testing the Equality of Two Means**

Two express mail services are competing for a contract with your company. You tested the two services by sending 15 pieces of mail through each service. The delivery times are as follows:

Service A			Service B		
14.3	15.9	17.5	15.6	16.5	17.8
14.4	16.2	18.1	15.8	17	17.9
15.5	16.7	18.2	16	17.1	18.3
15.9	16.8	18.5	16.1	17.2	18.4
15.9	16.9	18.6	16.3	17.3	18.7

Do a 5% test to determine if there are different delivery times between Service A and Service B.

Solution: $R_A = 153$

$$U_A = n_A n_B + \frac{n_A(n_A+1)}{2} - R_A = 15(15) + \frac{15(15+1)}{2} - 153 = 192$$

$$\mu_u = \frac{n_A n_B}{2} = \frac{15(15)}{2} = 112.5$$

$$\sigma_u = \sqrt{\frac{n_A n_B (n_A + n_B + 1)}{12}} = \sqrt{\frac{15(15)(15 + 15 + 1)}{12}} = 24.11$$

$$\frac{U_A - \mu_A}{\sigma_u} = \frac{192 - 112.5}{24.11} = 3.3 > 1.96 = z_{.05/2}$$

So there is enough evidence to conclude that the delivery times are different.

Example 5 **Kruskal-Wallis Test**

Thirty economists from different backgrounds are asked to estimate the unemployment rate in the next quarter. The unemployment estimates are ranked in the following table.

Academic Economist	Private Economist	Government Economist
1	3	4
10	2	5
9	8	15
6	7	16
11	13	17
12	14	18
20	19	21
22	24	23
28	26	25
29	30	27

Can you argue that the economists working in different professions do not have the same average estimate of the unemployment rate? Use a 5% level of significance.

Solution: $R_1 = 148$, $R_2 = 146$, $R_3 = 171$

$$K = \frac{12}{n(n+1)}[R_1^2/n_1 + R_2^2/n_2 + R_3^2/n_3] - 3(n+1)$$

$$= \frac{12}{30(30+1)}[148^2/10 + 146^2/10 + 171^2/10] - 3(30+1)$$

$$= 0.498 < \chi^2_{2,5\%} = 5.99$$

We do not have enough evidence to reject the null hypothesis.

Example 6 **Spearman's Rank-Correlation Test**

A consumer organization wants to know if consumers receive their money's worth when purchasing a stereo. They sampled ten stereos and ranked the quality and price of each set. A higher rank indicates a better product and a higher price.

Stereo	(1) Price Rank	(2) Quality Rank	d (1) – (2)	d^2
1	1	1	0	0
2	8	3	5	25
3	2	2	0	0
4	3	4	–1	1
5	7	5	2	4
6	4	6	–2	4
7	6	7	–1	1
8	5	8	–3	9
9	9	9	0	0
10	10	10	0	0
Total				44

Use a 5% level of significance to determine if consumers get what they pay for.

Solution: If consumers get what they pay for, the quality and price ranks should have a positive correlation.

H_0: $\rho = 0$
H_1: $\rho > 0$

$\Sigma d^2 = 44$

$$r_s = 1 - \frac{6 \times \Sigma d^2}{n(n^2-1)} = 1 - \frac{6(44)}{10\,(10^2-1)} = 0.73$$

$$t = \frac{r_s - \rho}{\sqrt{\frac{1-r_s^2}{n-2}}} = \frac{0.73-0}{\sqrt{\frac{1-0.73^2}{10-2}}} = 3.02 > 1.86 = t_{5\%,8}$$

So we reject the null hypothesis and conclude that consumers do get their money's worth.

Example 7 **Runs Test**

In a conference, forty economists were asked if they think we are out of the recession. The answering pattern, in order, is

y y n n y n y y y n n y y n y y n n n n y y n y y n y n n n y y n n y n y n y n

where y indicates a yes vote and n indicates a no vote.

The person conducting this survey suspects that an economist's answer tends to be affected by the answer just preceding it. Can you verify this? Do a 5% test.

Solution: If an economist's answer is affected by the previous answer, then the answering pattern should exhibit momentum.

H_0: no momentum
H_1: momentum

Runs = 24

$$\mu = \frac{2n_1n_2}{n} + 1 = \frac{2(20)(20)}{40} + 1 = 21$$

$$\sigma_R = \sqrt{\frac{2n_1n_2\,(2n_1n_2 - n)}{n^2(n-1)}} = \sqrt{\frac{2(20)(20)[2(20)(20)-40]}{40^2(40-1)}} = 4.47$$

$$\frac{R-\mu_R}{\sigma_R} = \frac{24-21}{4.47} = 0.67 < 1.64$$

So we cannot reject the null hypothesis.

Supplementary Exercises

Multiple Choice

1. Count the runs in the following data: + – + – – – – –
   a. 4
   b. 3
   c. 8
   d. 5
   e. 2

2. In a nonparametric test, it is assumed that
   a. the data are normally distributed.
   b. the test statistic is not normally distributed.
   c. the data are non-normal.
   d. the test statistic is normally distributed.
   e. the data may or may not be normally distributed

3. In conducting a statistical test, the research assistant messed up the order of the data and the test result was seriously biased. What kind of test was being done?
   a. sign test.
   b Mann-Whitney *U* test.
   c. Spearman's rank correlation test.
   d. runs test.
   e. Kruskal-Wallis test.

4. Which of the following statistics are not constructed on the ranks of the data?
   a. the sign test.
   b. the Mann-Whitney *U* test.
   c. the Wilcoxon signed-rank test.
   d. Spearman's rank correlation test.
   e. the Kruskal-Wallis test.

5. Which of the following tests is used to compare three or more means?
   a. the sign test.
   b. Mann-Whitney *U* test.
   c. the Wilcoxon signed-rank test.
   d. Spearman's rank correlation test.
   e. Kruskal-Wallis test.

6. Which of the following tests can compare two means of two groups with different sample sizes?
   a. the sign test.
   b. the Mann-Whitney *U* test.
   c. the Wilcoxon signed-rank test.
   d. Spearman's rank correlation test.
   e. Kruskal-Wallis test.

7. Which of the following tests uses the quantitative difference between data to conduct statistical inference?
   a. the sign test.
   b. the Mann-Whitney *U* test.
   c. the Wilcoxon signed-rank test.
   d. Spearman's rank correlation test.
   e. Kruskal-Wallis test.

8. If you are interested in seeing if a basketball player is a streaky shooter (tends to make many baskets in a row) you would use
   a. a runs test.
   b. a Mann-Whitney *U* test.
   c. a Wilcoxon signed-rank test.
   d. Spearman's rank correlation test.
   e. Kruskal-Wallis test.

9. The expected number of runs in n flips of a coin are
   a. n
   b. $n - 1$
   c. $2n_1n_2/n$
   d. $2n_1n_2/(n + 1)$
   e. $[2n_1n_2/n] + 1$

10. Suppose you ranked four students on the basis of their SAT scores (from 1 to 4, 1 being the highest SAT score). If these students' SAT scores have a rank correlation of –1 with the rank of their high school grades, then the rankings for high school grades would imply that
    a. the student with the highest SAT score also had the highest grade.
    b. the student with the highest SAT score had the lowest grade.
    c. the student with the lowest SAT score also had the lowest grade.
    d. the student with the highest SAT score had the middle grade.
    e. cannot be determined by the information given.

11. Suppose you ranked four students on the basis of their SAT scores (from 1 to 4, 1 being the highest SAT score). If these students' SAT scores have a rank correlation of +1 with the rank of their high school grades, then
    a. the student with the lowest SAT score had the highest grades.
    b. the student with the lowest SAT score also had the lowest grades.
    c. the student with the highest SAT score had the lowest grades.
    d. the student with the highest SAT score had the middle grade.
    e. the rankings for high school cannot be determined by the information given.

12. If we are interested in testing whether the median SAT score for Hannahtown high school is 1000, we would use
    a. a runs test.
    b. a Mann-Whitney *U* test.
    c. a Wilcoxon signed-rank test.
    d. Spearman's rank correlation test.
    e. sign test.

13. Count the number of runs in the following 6 flips of a coin: H T H H T T
    a. 6
    b. 5
    c. 4
    d. 3
    e. 2

14. If we are interested in testing whether the forecasts for economic growth made by government and academic economists differ from each other, we can use
    a. a runs test.
    b. a Mann-Whitney *U* test.
    c. a Wilcoxon signed-rank test.
    d. Spearman's rank correlation test.
    e. a sign test.

True/False (if false, explain why)

1. Nonparametric tests get this name because they do not use test statistics such as the t-test.
2. Many nonparametric tests use a ranking method to test hypotheses.
3. The Kruskal-Wallis test is used to test the equality of three or more means when the populations are not normal.
4. A rank-sum test can be used to test the hypothesis that changes in stock prices are random.
5. Nonparametric tests assume that the data are normally distributed.
6. In the runs test, the data follow a normal distribution.
7. Spearman's rank correlation measures the correlation between the rankings of two variables.
8. The Wilcoxon signed-rank test is used to compare the difference between two groups of data when the data are paired.
9. The sign test is used to compare the difference between two groups of data when the data are paired.
10. The Spearman's rank correlation has a maximum value of 1 and a minimum value of 0.
11. If we want to test whether data carry some kind of momentum, we may use the sign test.
12. If we are testing the null hypothesis that data does not carry some kind of momentum, we use the runs test and put the rejection region in the left tail.
13. In conducting the Kruskal-Wallis test, we should make sure that there is the same amount of data in each group.
14. In conducting the Mann-Whitney *U* test, it is necessary that we know the rank sum of each group.

Questions and Problems

1. It is widely believed that a student's mathematical ability can help his or her science grade. A mathematics teacher collected the grades of ten students and ranked the grades as follows:

Rank of math grade	1	2	3	4	5	6	7	8	9	10
Rank of science grade	2	4	1	5	9	3	6	10	8	7

   Do the above data support the hypothesis that the two grades are positively correlated? Use a 5% level of significance.

2. Twenty-five graduates who majored in computer science were asked to give their starting salaries. Using the sign test, can you reject the hypothesis that the median salary is $30 thousand? Do a 5% test.

Salaries (in thousands of dollars)				
35.3	34.3	26.1	28.2	28.9
32.8	37.2	32.8	31.6	32.8
31.2	29.7	26.3	31.3	32.7
32.8	33.8	32.1	32.3	35.4
29.7	31.6	38.6	31.8	31.2

3. Fifty graduates, 25 males and 25 females, who majored in computer science were asked to give their starting salaries. The results are given below in thousands of dollars.

Males					Females				
25.1	25.2	25.3	25.8	25.9	24.3	24.5	25.7	26.0	26.2
26.3	26.4	26.8	26.9	27.0	26.5	26.7	27.3	27.6	27.9
27.1	27.2	27.4	27.5	27.8	28.1	28.3	28.3	28.4	28.5
27.9	28.1	28.3	28.7	28.8	28.6	28.6	28.7	28.9	28.9
29.3	29.4	29.5	29.7	30.5	29.1	29.2	29.2	29.2	29.2

Use a 5% level of significance to determine if males receive a higher average salary than females.

4. A baseball player's last 40 at bats are recorded as follows. (H is a hit and N is no hit.)

H N H H H N H H H H N N N N N N N H H H N N N N N N N N N N H H N N N N H N N N

Is this batter a streaky hitter? Use a 5% level of significance.

5. Thirty-five workers are divided into three groups. The first group works in a quiet environment. The second group works while playing classical music. The third group works while playing easy listening music. The productivities of the three groups are given below.

Group 1	1	7	9	12	13	15	16	22	27	21	30	33	35
Group 2	2	4	5	6	11	17	18	19	20	25	28	29	
Group 3	3	8	10	14	23	24	26	31	32	34			

Do a 5% test to determine if the three groups have the same average productivity.

6. Use the data from Problem 5 to compare the first and second groups. Can you argue that the average productivities are different? Do a 5% test.

Answers to Supplementary Exercises

Multiple Choice

1. a	6. b	11. b
2. c	7. c	12. e
3. d	8. a	13. c
4. a	9. e	14. c
5. e	10. b	

True/False

1. False. They get their name because the distribution of the data is not known.
2. True
3. True
4. False. Use a runs test.
5. False. Data do not necessarily have to be normally distributed.
6. False. Data are dichotomous, so they cannot be normally distributed.
7. True
8. True
9. True
10. False. Minimum value is –1.
11. False. Use a runs test.
12. True
13. False. Amount of data can differ.
14. False. If we know the rank sum of one group we can conduct the test.

Questions and Problems

1. H_0: $\rho \le 0$, not positively correlated.
   H_1: $\rho > 0$, positively correlated.

$$r_s = 1 - \frac{6(50)}{10(10^2 - 1)} = .0697$$

$$t = \frac{0.697 - 0}{\sqrt{\frac{1 - 0.697^2}{10 - 2}}} = 2.75 > 1.86$$

Reject the null hypothesis. There is positive correlation.

2. H_0: $p \leq 1/2$
   H_1: $p > 1/2$

   p = the percentage of numbers lower than 30.

$$z = \frac{19/25 - 1/2}{\sqrt{\frac{1/2(1 - 1/2)}{25}}} = 2.6 > 1.64$$

   Reject the null hypothesis.

3. The rankings of the males and females are:

Males					Females				
3	4	5	7	8	1	2	6	9	10
11	12	15	16	17	13	14	20	23	25.5
10	19	21	22	24	27.5	30	30	32	33
25.5	27.5	30	36.5	38	34.5	34.5	36.5	39.5	39.5
46	47	48	49	50	41	43.5	43.5	43.5	43.5

   Rank sum of males, $R_1 = 599.5$.

$$U_1 = 25(25) + \frac{25(25+1)}{2} - 599.5 = 350.5$$

$$\mu_U = \frac{25(25)}{2} = 312.5$$

$$\sigma_u = \sqrt{\frac{(25(25)(25+25+1)}{12}} = 51.54$$

$$z = \frac{350.5 - 312.5}{51.54} = 0.74 < 1.64$$

So we cannot reject the null hypothesis.

4. H_0: not a streaky hitter.
   H_1: a streaky hitter.

   We can test the null hypothesis by examining the number of runs. If the hitter is streaky, there will be fewer runs than we would expect because both the hits and outs will come in bunches.

   $R = 12, \; n_1 = 14, \; n_2 = 26$

$$\mu_R = \frac{2(14)(26)}{40} + 1 = 19.2$$

$$\sigma_R = \sqrt{\frac{2(14)(16)[2(14)(16) - 40]}{40^2 \; (40-1)}} = 1.71$$

$$z = \frac{12 - 19.2}{1.71} = -4.21 < -1.64$$

   We reject the null hypothesis. The hitter is a streak hitter.

5. H_0: equally productive.
   H_1: not equally productive.

$$\chi^2 = \frac{12}{35(35+1)}\left(\frac{241^2}{13} + \frac{184^2}{12} + \frac{205^2}{10}\right) - 3(35+1) = 1.44 < 5.99 = \chi^2_{.05}$$

   Cannot reject the null hypothesis.

6. To answer this question, we need to re-rank the data in groups 1 and 2.

H_0: equally productive.
H_1: not equally productive.

$R_1 = 185$

$$U_1 = 13(12) + \frac{13(13+1)}{2} - 185 = 62$$

$$\mu_u = \frac{13(13)}{2} = 78$$

$$\sigma_u = \sqrt{\frac{13(12)(13+12+1)}{12}} = 18.38$$

$$z = \frac{62-78}{18.38} = -0.87 > -1.96$$

Cannot reject the null hypothesis.

CHAPTER 18

TIME-SERIES: ANALYSIS, MODEL, AND FORECASTING

Chapter Intuition

Previously, you learned how different variables could be related to one another by using regression analysis. In most of the examples that were discussed, you were looking at ***cross-sectional data***. Cross-sectional data are data which occur at the same time, but which are examined across many different individuals or companies. For example, if we were interested in determining factors that influence department store sales in 2003, we could collect information such as sales, prices, and strength of the economy in 2003 for many different stores. Because we are determining sales in one time period (2003), across many stores, we are using cross-sectional data.

A different type of analysis that could be conducted is to look at sales over many different months or years for the same store. Because we are analyzing the data over different time periods, we are looking at time series data. While some of the same techniques used in cross-sectional analysis are also applicable to time series analysis, the use of time series data may necessitate the use of special techniques not required in cross-sectional analysis. For example, if we are using monthly sales data to find a relationship between the sales of Sears and various economic factors, we should also deal with possible seasonal effects which occur in retail sales (e.g., higher sales around Christmas). This chapter discusses how to deal with some of these problems and presents special techniques that can be used in time series analysis.

Chapter Review

1. ***Time series data*** are numbers that are recorded over time. ***Cross-sectional data*** are numbers that are recorded across different companies, industries, or individuals in the same time period.
2. When analyzing a time series, we sometimes find it easier to conduct the analysis if we decompose the time series into several observable components.

a. The ***trend component*** is the part of the time series that reflects permanent information in both the short run and the long run.

b. The ***seasonal component*** is the intrayear pattern, which constantly repeats itself from year to year. For example, the sales revenues of a toy store would be expected to experience higher sales during the Christmas shopping season.

c. The ***cyclical component*** consists of long-term oscillatory patterns that are unrelated to seasonal behavior. The U.S. Department of Commerce has specified a number of time series that are used to identify the peaks and valleys in the business cycle. ***Leading indicators*** get their name because they tend to have their turning points prior to turns in economic activity. ***Coincidence indicators*** tend to have their turning points at the same time as turns in economic activity. ***Lagging indicators*** tend to have their turning points following the turning points in economic activity.

d. The ***irregular component*** of a time series is the random pattern that may show up in either the long run or the short run.

3. One of the simplest approaches to analyzing a time series is to use a ***simple moving average***. In an n-period simple moving average we just find the mean of the series for the previous n periods. For example, we might look at the mean value of the Dow Jones Industrial Average over the last 30 days.

4. ***Exponential smoothing*** is a method that is commonly used in forecasting in which a weighted average of some past values of the variable are used to forecast values of the variable.

5. In a ***time trend regression***, we use a time trend as our independent variable. The time trend can be either linear or nonlinear. In a nonlinear time trend, we use a quadratic model to specify the trend relationship in order to pick up any nonlinear effects.

6. The ***Holt-Winters forecasting model*** consists of both an exponentially smoothed component and a trend component to forecast future values.

7. In an ***autoregressive process***, current values of the dependent variable are assumed to depend on its past values.

Useful Formulas

k-term moving average:

$$Z_t = \frac{1}{k}\sum_{i=0}^{k-1} x_{t-i}, \qquad t = (k, \ldots, N)$$

Exponential smoothing:

$$S_{t+1} = \alpha X_t + (1-\alpha)S_t$$

Linear time trend:

$$X_{it} = \alpha_i + \beta_i t + \varepsilon_{it}$$

Nonlinear time trend:

$$X_{it} = \alpha_i + \beta_i t + \gamma_i t^2 + \varepsilon_{it}$$

$$X_{it} = \alpha_i + \beta_i t + \gamma_i t^2 + \delta_i t^3 + \varepsilon_{it}$$

k-th order autoregressive process:

$$y_t = \alpha + \phi_1 y_{t-1} + \phi_2 y_{t-2} + \ldots + \phi_k y_{t-k} + \varepsilon_t$$

Example Problems

Example 1 **Identifying the Components of a Time Series**

The following data are for sales of sports coats at a department store over the last three years.

Year	Quarter	Sales
1	1	1,731
	2	935
	3	2,529
	4	2,400
2	1	1,789
	2	1,000
	3	2,931
	4	2,350
3	1	1,875
	2	1,150
	3	2,950
	4	2,650

Graph the data and identify the components of this time series.

Solution:

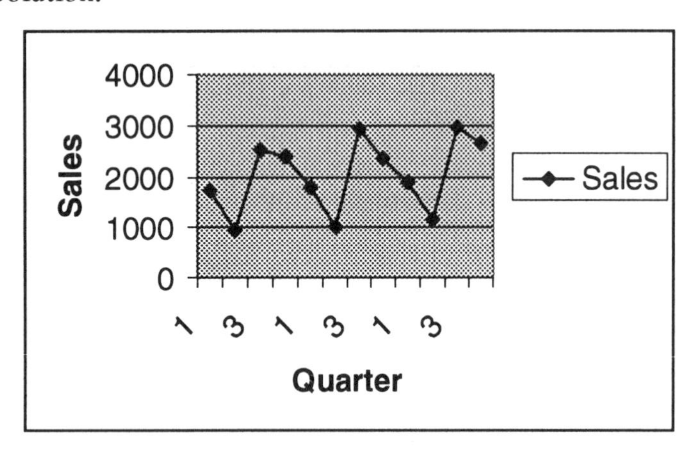

Example 2 **Simple Moving Average and Centered Moving Average**

Use the data given in Example 1 to find the four-period moving average and the centered moving average.

Solution: The first four-period moving average would be the arithmetic average of the first four observations.

Moving average = (1731 + 935 + 2529 + 2400)/4 = 1898.75

The moving average for the first four observations is centered around observations 1 to 4 or at observation 2.5. The second moving average is centered around observations 2 to 5 or at observation 3.5, and so on. In order to find the moving average at observation 3, we need to take the average of our moving averages for observation 2.5 and observation 3.5. This is called a centered moving average. The centered moving average for the third quarter of the first year is the average of the moving average at quarter 2.5 and the moving average at quarter 3.5.

Centered MA for Quarter 3 = (1898.75 + 1913.25)/2 = 1906

The rest of the moving averages and centered moving averages can be computed in a similar manner as follows:

Year	Quarter	Sales	4-Period MA	Centered MA
1	1	1731		
	2	935		
			1898.75	
	3	2529		1906
			1913.25	
	4	2400		1921.38
			1929.5	
2	1	1789		1979.75
			2030	
	2	1000		2023.75
			2017.5	
	3	2931		2028.25
			2039	
	4	2350		2057.75
			2075.5	
3	1	1875		2078.88
			2081.25	
	2	1150		2118.75
			2156.25	
	3	2950		
	4	2650		

Example 3 **Seasonal and Irregular Component**

Use your results from Example 2 to construct the seasonal and irregular components of the data.

Solution: The seasonal and irregular components are constructed by dividing the sales by the centered moving average and multiplying by 100.

The seasonal and irregular component for the third quarter of the first year would be

$$(2529/1906)100 = 132.69$$

Year	Quarter	(X_t) Sales	(Z_t) Centered MA	$(X_t/Z_t)100$ $S_t \times I_t$
1	1	1731		
	2	935		
	3	2529	1906	132.69
	4	2400	1921.38	124.91
2	1	1789	1979.75	90.36
	2	1000	2023.75	49.41
	3	2931	2028.25	144.51
	4	2350	2057.75	114.20
3	1	1875	2078.88	90.19
	2	1150	2118.75	54.28
	3	2950		
	4	2650		

Example 4 **Seasonal Component**

Use your results from Example 3 to separate the seasonal and irregular components of the time series.

Solution: Notice in Example 3, we have two seasonal and irregular components for each quarter. To remove the irregular component from the seasonal factor, we take the average of each quarter's values.

$S_1 = (90.36 + 90.19)/2 = 90.28$ (1st quarter adjustment)

$S_2 = (49.41 + 54.28)/2 = 51.84$ (2nd quarter adjustment)

$S_3 = (132.69 + 144.51)/2 = 138.60$ (3rd quarter adjustment)

$S_4 = (124.91 + 114.20)/2 = 119.56$ (4th quarter adjustment)

Example 5 **Seasonally Adjusted Sales**

Use your results from Example 4 to seasonally adjust the original sales data. Plot the seasonally adjusted sales and the original sales data. Has the adjustment removed the seasonal effect?

Solution: To seasonally adjust the time series, we divide the original sales data by the seasonal factor and multiply by 100. (Note: we will round the seasonal factors off to 90, 52, 138, and 120). The first year's sales in the first quarter would then be

Adjusted Sales = (1731/90)(100) = 1923.33

The rest of the seasonally adjusted sales would be found in a similar manner.

Year	Quarter	(X_t) Sales	(S_t) Seasonal Adjustment	$(X_t/S_t)100$ Adjusted Sales
1	1	1731	90	1923.33
	2	935	52	1798.08
	3	2529	138	1832.61
	4	2400	120	2000.00
2	1	1789	90	1987.78
	2	1000	52	1923.08
	3	2931	138	2123.91
	4	2350	120	1958.33
3	1	1875	90	2083.33
	2	1150	52	2211.54
	3	2950	138	2137.68
	4	2650	120	2208.33

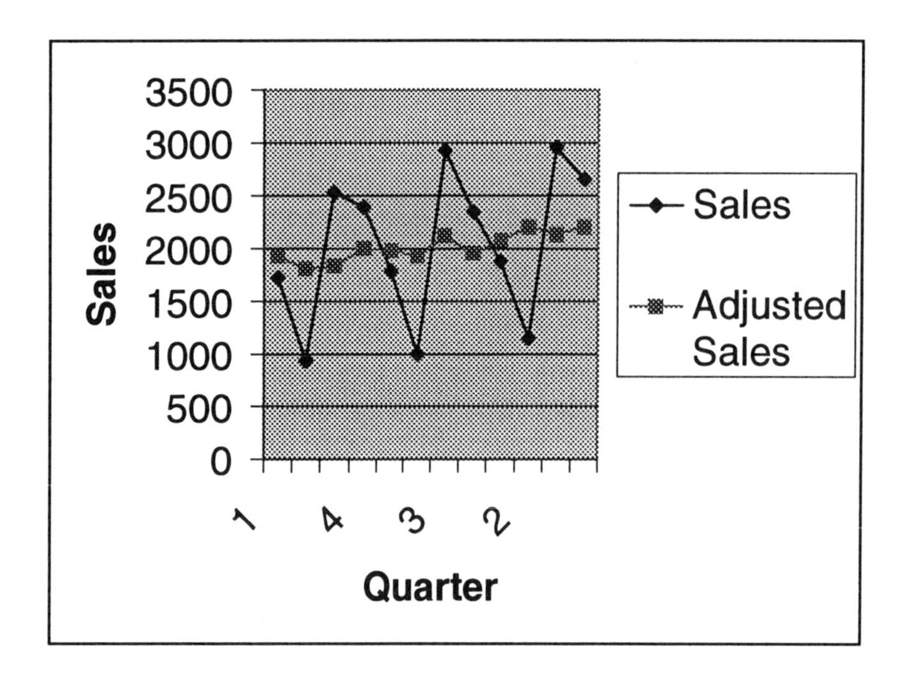

Example 6 **Linear Trend Regression**

Run a regression on the deseasonalized data you computed in Example 5 to estimate the trend line. Use the trend line to forecast sales for the next four quarters.

Solution:

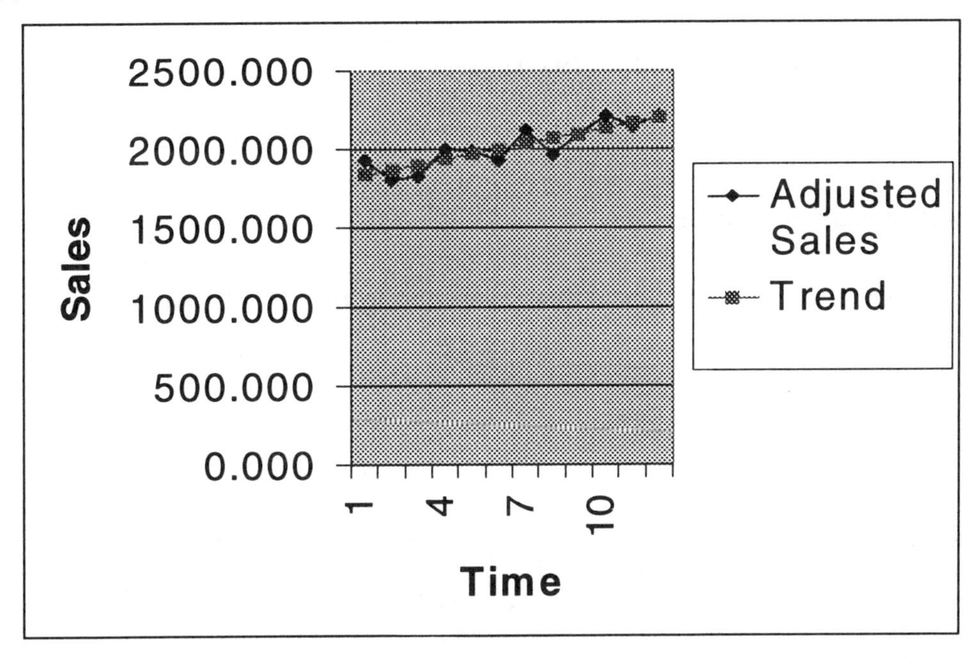

To estimate the trend line, we regress the seasonally adjusted data on time (t = 1, 2, . . ., 12). The estimated regression is

$$\text{Adjusted sales}_t = 1802.64 + 32.77\,t, \quad R^2 = .735$$

To forecast adjusted sales in periods 13-15, we substitute the time period into the regression

$$\text{Adjusted sales}_{13} = 1802.64 + 32.77(13) = 2228.69$$

$$\text{Adjusted sales}_{14} = 1802.64 + 32.77(14) = 2261.46$$

$$\text{Adjusted sales}_{15} = 1802.64 + 32.77(15) = 2294.24$$

$$\text{Adjusted sales}_{16} = 1802.64 + 32.77(16) = 2327.01$$

From the above graph, we can see that the trend line forecasts seasonally adjusted sales. To get a forecast of the sales we need to make an adjustment using the seasonal index we computed. The adjustment is to multiply the seasonal index and divide by 100.

$$\text{Sales}_{13} = 2228.69(90)/100 = 2005.82$$

$$\text{Sales}_{14} = 2261.46(52)/100 = 1175.96$$

$Sales_{15} = 2294.24(138)/100 = 3166.05$

$Sales_{16} = 2327.01(120)/100 = 2792.41$

Example 7 **Exponential Smoothing**

The following are car sales data at a local dealership during the first 12 weeks of the year.

Week	Sales	Week	Sales
1	15	7	11
2	14	8	10
3	13	9	12
4	12	10	10
5	14	11	12
6	13	12	13

Use the exponential smoothing method to forecast the sales for the first four weeks. Assume $\alpha = 0.2$.

Solution: The formula for exponential smoothing looks at the last period's forecast, plus an adjustment based on our forecast error from the previous period.

$$S_t = S_{t-1} + \alpha(X_{t-1} - S_{t-1})$$

$S_1 = 15$

$S_2 = 15 + .2(15 - 15) = 15$

$S_3 = 15 + .2(14 - 15) = 14.8$

$S_4 = 14.8 + .2(13 - 14.8) = 14.44$

Example 8 **Autoregressive Model**

Use the data given in Example 1 to estimate the parameters of a first-order autoregressive model. Forecast sales one period into the future using this model.

Solution: To estimate the parameters of a first-order autoregressive model, we regress the sales against last period's sales. The regression we estimate is

$$X_t = \alpha + \beta X_{t-1} + \varepsilon_t$$

Using any regression package, we can estimate the model as

$$\hat{X}_t = 2292.81 - .123 X_{t-1} \quad R^2 = .0142$$

We can forecast the next quarter's sales by substituting the current period's sales into the estimated regression to get.

$$\hat{X}_t = 2292.81 - .123(2650) = 1966.86$$

Example 9 **Exponential Smoothing - Choosing the Smoothing Parameter**

Suppose the smoothing parameter α is unknown. Suggest a method for estimating α. Use this method to find the best α using the data from Example 7.

Solution: One method for selecting the best smoothing parameter is to choose the α with the smallest mean squared error (MSE) for the forecasts. The MSE is computed as

$$\text{MSE} = \Sigma e_t^2 / n$$

where n is the number of forecasts.

The MSE for different values of α are presented below.

α	MSE	α	MSE
.1	5.00	.6	2.177
.2	3.32	.7	2.175
.3	2.65	.8	2.20
.4	2.36	.9	2.27
.5	2.23		

So the best α is .7.

Example 10 **Holt-Winters Forecasting Method**

Use the Holt-Winters method to forecast the sales in the 10th period for the following data. Assume that the smoothing constants are $\alpha = 0.6$ and $\beta = 0.5$.

Time	Sales	Time	Sales
1	37.8	6	43.4
2	38.0	7	43.9
3	38.6	8	44.7
4	40.3	9	45.7
5	41.2	10	46.4

Solution: $S_t = .6X_t + (1 - .6)(S_{t-1} + T_{t-1})$

$T_t = .5(S_t - S_{t-1}) + (1 - .5)T_{t-1}$

To begin the system we use $T_1 = 0$ and $S_1 = X_1 = 37.8$

$$\begin{aligned} S_2 &= \alpha X_2 + (1 - \alpha)(S_1 + T_1) \\ &= .6(38.0) + (1 - .6)(37.8 + 0) \\ &= 37.92 \end{aligned}$$

$$\begin{aligned} T_2 &= \beta(S_2 - S_1) + (1 - \beta)T_1 \\ &= .5(37.92 - 37.8) + (1 - .5)0 \\ &= .06 \end{aligned}$$

$$\begin{aligned} S_3 &= \alpha X_3 + (1 - \alpha)(S_2 + T_2) \\ &= .6(38.6) + (1 - .6)(37.92 + .06) \\ &= 38.352 \end{aligned}$$

$$\begin{aligned} T_3 &= \beta(S_3 - S_2) + (1 - \beta)T_2 \\ &= .5(38.352 - 37.92) + (1 - .5).06 \\ &= .246 \end{aligned}$$

The rest of the values are presented in the table below.

Time	Sales	S_t	T_t
1	37.8	37.8	0
2	38.0	37.92	.06
3	38.6	38.35	.246
4	40.3	39.62	.757
5	41.2	40.87	1.004
6	43.4	42.79	1.462
7	43.9	44.04	1.356
8	44.7	44.98	1.147
9	45.7	45.87	1.019
10	46.4	46.78	.962

Example 11 **Autoregressive Model**

Estimate the first-order autoregressive model for the data given below. Use the model to produce a forecast for the 16th period.

Time	X_t	Time	X_t
1	.50	9	1.05
2	1.14	10	1.06
3	1.01	11	1.14
4	1.11	12	1.30
5	1.44	13	.93
6	1.02	14	1.09
7	1.18	15	1.45
8	1.71		

The first-order autoregressive model is computed by regressing the current period value of *X* against the previous period. The estimated regression is

$$X_t = 1.305 - .106X_{t-1} \quad R^2 = 0.16$$

To forecast the value in the 16th period we substitute into the estimated regression to get

$$X_{16} = 1.305 - .106(1.45) = 1.151$$

Supplementary Exercises

Multiple Choice

1. To construct the current 30-day moving average of stock prices, we
   a. use the first 30 days of stock prices.
   b. use the most recent 30 days of stock prices.
   c. use any 30 days of stock prices.
   d. divide the most recent stock price by 30.
   e. multiply the most recent stock price by 30.

2. Time series analysis is used to analyze data
   a. over different time periods.
   b. across different companies.
   c. across different companies and across different time periods.
   d. that are qualitative.
   e. provided they are based on stock prices.

3. A record store notices that sales increase on Elvis's birthday. The "Elvis phenomenon" is
   a. a trend component.
   b. an irregular component.
   c. a seasonal component.
   d. a cyclical component.
   e. a random component.

4. If the last eight observations of a time series are: a, b, c, d, e, f, g, h, then the current three-period moving average would be
   a. $(a + b + c)/3$
   b. $(a + b + c) \times 3$
   c. $h/3$
   d. $(f + g + h)/3$
   e. $(f + g + h) \times 3$

5. Which of the following would likely be a seasonal component of a time series?
   a. holidays
   b. law suits
   c. recessions
   d. population growth
   e. depression

6. Which of the following would likely be a trend component of a time series?
   a. holidays
   b. law suits
   c. recessions
   d. population growth
   e. depression

7. Exponential smoothing requires
   a. past values of the time series.
   b. current values of the time series.
   c. both past and current values of the time series.
   d. estimation of a time trend regression.
   e. seasonal adjustments.

8. When using simple exponential smoothing, the use of a larger smoothing coefficient means
   a. more weight is given to past observations.
   b. less weight is given to current observations.
   c. more weight is given to current observations.
   d. equal weight is given to past and current observations.
   e. it is impossible to determine which observations receive greater weight.

9. The Holt-Winters forecasting model
   a. is a simple exponential smoothing model.
   b. is a time trend model.
   c. is a moving average model.
   d. contains both a time trend and exponential smoothing.
   e. is a cyclical model.

10. An autoregressive model
    a. is a time trend model.
    b. is a method for seasonally adjusting data.
    c. uses past values of the data to forecast future values.
    d. is an exponential smoothing model.
    e. is a cyclical model.

11. Variables that tend to precede turning points in economic activity are called
    a. leading indicators.
    b. coincidence indicators.
    c. lagging indicators.
    d. moving averages.
    e. exponentially smoothed values.

12. Variables that tend to match turning points in economic activity are called
    a. leading indicators.
    b. coincidence indicators.
    c. lagging indicators.
    d. moving averages.
    e. exponentially smoothed values.

13. Variables that tend to follow turning points in economic activity are called
    a. leading indicators.
    b. coincidence indicators.
    c. lagging indicators.
    d. moving averages.
    e. exponentially smoothed values.

14. Data can be deseasonalized by using
    a. the moving average.
    b. exponential smoothing.
    c. an autoregressive process.
    d. a trend regression.
    e. leading indicators.

15. If the sales for a company exhibit constant growth over time, the best method for forecasting would be
    a. a linear time trend.
    b. a log-linear time trend.
    c. a simple moving average.
    d. a first-order autoregressive model.
    e. a centered moving average.

True/False (if false, explain why)

1. A simple moving average could lie consistently above the original data.
2. A simple moving average will exhibit more variability than the original data.
3. A simple moving average can be used to forecast cross-sectional data.
4. Regression cannot be used to forecast a time series.
5. The more stable the time series, the smaller the smoothing parameter should be.
6. If a company's earnings exhibit high earnings in the first and third quarters and lower earnings in second and fourth quarters, the time series is cyclical.
7. The larger the smoothing coefficient in exponential smoothing, the greater the weight given to current observations.
8. Moving averages can be used to deseasonalize the data.
9. For a trend to exist, the data must be growing steadily over time.
10. The high sales that most stores experience around the holiday season are a cyclical effect.
11. Cyclical patterns are long-term oscillatory patterns that are related to seasonal behavior.
12. Leading economic indicators get their name because they are the most important indicators of economic growth.
13. When we are looking at quarterly data, a four-period simple moving average will exhibit a pronounced seasonal effect.

14. A simple time trend regression is the best method for forecasting a variable that shows constant growth over time.
15. The Holt-Winters model uses both smoothing and a trend to forecast a time series.

Questions and Problems

1. Briefly explain why we must deseasonalize data before we can use a trend to forecast future values.
2. Briefly explain why we cannot use exponential smoothing when a trend exists.
3. You are given the following quarterly sales information for XYZ Company over the last 3 years.

Year	Quarter	Sales
1	1	3,800
	2	3,827
	3	3,995
	4	3,525
2	1	4,010
	2	3,950
	3	4,102
	4	3,995
3	1	4,210
	2	4,101
	3	4,331
	4	4,459

Find the current four-period moving average.

4. Below are sales data for XYZ Company.

Time	Sales	Time	Sales
1	44.3	6	50.2
2	47.8	7	51.1
3	47.6	8	50.9
4	48.3	9	52.2
5	49.4	10	53.3

Use the Holt-Winters forecasting model to forecast sales for the 10 periods. Assume that $\alpha = 0.6$ and $\beta = 0.5$.

5. Below are sales data for ABC Company.

Period	Sales	Period	Sales
1	146.5540	16	176.6588
2	159.4821	17	173.3224
3	171.8967	18	170.7909
4	172.9721	19	169.0809
5	176.6022	20	167.7419
6	172.0441	21	172.9483
7	171.3946	22	178.0765
8	169.2607	23	177.8793
9	175.6464	24	173.5440
10	177.4715	25	176.0660
11	176.3656	26	179.8407
12	180.3656	27	173.2309
13	175.0408	28	170.4035
14	177.9652	29	169.2317
15	173.1053		

Use a computer program to estimate the first-order autoregressive model to forecast sales in the 30th period.

6. Below are the sales data for the Smart Computer Company over the last 20 months. Use a time trend regression to forecast sales for the 21st month.

Month	Sales	Month	Sales
1	370.2467	11	370.9667
2	370.8606	12	374.5309
3	371.2098	13	376.6496
4	372.3486	14	376.9085
5	372.7624	15	376.071
6	373.7477	16	378.6685
7	372.0169	17	376.0601
8	373.8907	18	370.9929
9	370.3686	19	379.3183
10	373.3517	20	378.1466

Answers to Supplementary Exercises

Multiple Choice

1. b
2. a
3. c
4. d
5. a
6. d
7. c
8. c
9. d
10. c
11. a
12. b
13. c
14. a
15. b

True/False

1. True
2. False. The moving average will be less variable.
3. False. A simple moving average is used to forecast time series data.
4. False. Regression analysis can be very useful for forecasting time series data.
5. False. If a time series is stable, we would want to place more weight on the most recent observation, so we should use a larger smoothing coefficient.

6. False. Seasonal effect.
7. True
8. True
9. False. The data could also be falling over time.
10. False. Seasonal effect.
11. False. The cyclical component should be unrelated to seasonal effects.
12. False. Get their name because they precede turning points in economic activity.
13. False. Because each moving average observation incorporates a full year of data, it will not show a seasonal effect.
14. False. A log-linear time trend is best for forecasting a time series that shows constant growth.
15. True

Questions and Problems

1. Because a simple time trend cannot capture the seasonal effect, we get more accurate forecasts if we remove the seasonal effect from the time series. Once the deseasonalized data have been forecast using a trend, we can adjust the values for the seasonal effect.
2. Exponential smoothing should not be used when a trend exists because the forecasts will always lag the actual values of the time series.
3. Moving average = (4,210 + 4,101 + 4,331 + 4,459)/4 = 4,275.25
4.

Time	X	S	T	Forecast
1	44.3	44.3	0	44.3
2	47.8	46.4	1.05	47.45
3	47.6	47.54	1.095	48.635
4	48.3	48.434	0.9945	49.4285
5	49.4	49.411	0.98595	50.39735
6	50.2	50.27894	0.926745	51.20569
7	51.1	51.14227	0.89504	52.03731
8	50.9	51.35493	0.553845	51.90877
9	52.2	52.08351	0.641214	52.72472
10	53.3	53.06989	0.813797	53.88369

5. To estimate a first-order autoregressive model, we regress sales on sales lagged one period. Using a computer program the estimated model is

$$X_t = 96.38 + .447\ X_{t-1}$$

To forecast sales in the 30th period we substitute in sales in period 29 to get

$$X_{30} = 96.38 + .447(169.2317) = 172.03$$

$$R^2 = .466$$

6. Using a computer program we estimate the time trend regression as

$$X_t = 369.22 + .4389\ t$$

$$R^2 = .700$$

The forecasted sales in period 21 are

$$X_{21} = 369.22 + .4382(21) = 378.44$$

CHAPTER 19

INDEX NUMBERS AND STOCK MARKET INDEXES

Chapter Intuition

Throughout the text, we have learned that one of the most important uses of statistics is to summarize data and to make comparisons. Often we are interested in comparing the past and the present. For example, we may be interested in how prices have changed over time. However, simply looking at changes in the price of one good at a time is impractical and not very useful. Although the prices of most items have risen over the past 10 years, the cost of many electronic goods such as computers, VCRs, and televisions have fallen dramatically. In addition, many items commonly used today were not even in existence several years ago. One way to compare price changes in two different time periods is by constructing an index of prices. This chapter explains the different methods for constructing index numbers.

We can also use index numbers to make comparisons in the same time period. For example, many people are interested in how their stocks have done compared to the rest of the stock market. One way to make this comparison is to construct an index of stock prices such as the Dow Jones Industrial Average and to compare the performance of our stocks to the index.

Chapter Review

1. Simply put, a ***price index*** represents a weighted average cost for purchasing a bundle of goods. By comparing the price index for a bundle of goods in different periods, we can see if it has gotten more or less expensive to purchase these goods.

2. One difficulty with constructing a price index is determining the appropriate bundle of goods. Although food makes an important part of a household's budget, and therefore should be represented in our price index, changes in eating habits mean that we are consuming different foods than previous generations. In addition, there are goods that are an important part of most household budgets today, such as VCRs and computers that did not even exist 20 years ago.

3. In order to construct an index, we need to determine the ***base year***. The base year is the reference point from which we measure changes in prices, quantities, or values.
4. The simplest price index is a ***price relative***. A price relative is the ratio of the current price of one commodity to its base year price. Price relatives allow us to look at the price change of a single commodity between the base year and the current year.
5. A ***simple aggregate price index*** is the ratio of the total cost of various commodities in the current year to the total cost in the base year. There are two problems with this type of index: it does not consider the relative importance of each of the commodities, and the units used can affect the index value. For example, if we use the price of one egg, rather than the cost of one dozen eggs, when we construct the index, eggs will become less important in the value of the index.
6. ***Weighted relative price indexes*** look at both the prices and quantities. The ***Laspeyres price index*** uses base year quantities. The ***Paasche price index*** uses current year quantities. ***Fisher's ideal price index*** is a geometric average of the Laspeyres and Paasche indexes.
7. If we are interested in how the quantity of goods purchased has changed, we can construct a ***quantity index***.
8. If we are interested in how the value of goods purchased has changed, we can construct a ***value index***.

Useful Formulas

Price Indexes

Price Relatives

$$P_{it} / P_{i0}$$

Simple Aggregate Price Index

$$I_t = \frac{\sum_{i=1}^{n} P_{it}}{\sum_{i=1}^{n} P_{i0}}$$

Simple Relative Price Index

$$I_t = \frac{\sum_{i=1}^{n} P_{it} / P_{i0}}{n} \times 100$$

Laspeyres Index

$$I_t = \frac{\sum_{i=1}^{n} p_{it} q_{i0}}{\sum_{i=1}^{n} p_{i0} q_{i0}} \times 100$$

Paasche Index

$$I_t = \frac{\sum_{i=1}^{n} p_{it} q_{it}}{\sum_{i=1}^{n} p_{i0} q_{it}} \times 100$$

Fisher's Ideal Price Index

$$FI_t = \sqrt{\frac{\sum p_{it} q_{it}}{\sum p_{i0} q_{it}} \times \frac{\sum p_{it} q_{i0}}{\sum p_{i0} q_{i0}}} \times 100$$

Quantity Indexes

Laspeyres

$$I_t = \frac{\sum_{i=1}^{n} q_{it} p_{i0}}{\sum_{i=1}^{n} q_{i0} p_{i0}} \times 100$$

Paasche

$$I_t = \frac{\sum_{i=1}^{n} q_{it} p_{it}}{\sum_{i=1}^{n} q_{i0} p_{it}} \times 100$$

Fisher's Quantity Index

$$FI_t = \sqrt{\frac{\sum q_{it} p_{it}}{\sum q_{i0} p_{it}} \times \frac{\sum q_{it} p_{i0}}{\sum q_{i0} p_{i0}}} \times 100$$

Value Indexes

$$I_t = \frac{\sum_{i=1}^{n} q_{it} p_{it}}{\sum_{i=1}^{n} q_{i0} p_{i0}} \times 100$$

Example 1 **Price Relatives**

Below are the prices and quantities for three commodities in 1992 and 1993.

Commodity	1992		1993	
	Price	Quantity	Price	Quantity
Eggs	$1.00	22 dozen	$1.21	20 dozen
Milk	$1.95	16 gallon	$2.05	18 gallon
Beef	$3.22	11 lb.	$3.20	12 lb.

Find the price relative for each commodity using 1992 as the base year.

Solution:

Price relative = P_{ti} / P_{0i}

Eggs = 1.21/1.00 = 1.21

Milk = 2.05/1.95 = 1.05

Beef = 3.20/3.22 = .99

Example 2 **Simple Aggregate Price Index**

Using the data from Example 1, compute the simple aggregate price index. Use 1992 as the base year.

Solution:

$$I_t = \frac{\sum p_{ti}}{\sum p_{0i}} \times 100$$

$$= \frac{1.21 + 2.05 + 3.20}{1.00 + 1.95 + 3.22} \times 100 = 104.7$$

Example 3 **Simple Relative Price Index**

Use the data and your results from Example 1 to compute the simple relative price index.

Solution:

$$I_t = \frac{\Sigma p_{ti} / p_{0i}}{n} = \frac{1.21 + 1.05 + .994}{3} \text{ x } 100 = 108.47$$

Example 4 **Laspeyres Price Index**

Using the data from Example 1, compute the Laspeyres price index. Use 1992 as the base year.

Solution:

$$I_t = \frac{\Sigma p_{ti} \, q_{0i}}{\Sigma p_{0i} \, q_{0i}} \text{ x } 100$$

$$= \frac{1.21(22) + 2.05(16) + 3.20(11)}{1.00(22) + 1.95(16) + 3.22(11)} \text{ x } 100 = 106.77$$

Example 5 **Paasche Price Index**

Using the data from Example 1, compute the Paasche price index. Use 1992 as the base year.

Solution:

$$I_t = \frac{\sum p_{ti}\, q_{ti}}{\sum p_{0i}\, q_{ti}} \times 100$$

$$= \frac{1.21(20) + 2.05(18) + 3.20(12)}{1.00(20) + 1.95(18) + 3.22(12)} \times 100 = 106.14$$

Example 6 **Fisher's Ideal Price Index**

Use your results from Examples 4 and 5 to compute Fisher's ideal price index.

Solution:

$$FI_t = \sqrt{\frac{\sum p_{ti}\, q_{ti}}{\sum p_{0t}\, q_{ti}} \frac{\sum p_{ti}\, q_{0i}}{\sum p_{0i}\, q_{0i}}} \times 100$$

$$= \sqrt{(1.068)(1.061)} \times 100 = 106.4$$

Example 7 **Laspeyres Quantity Index**

Below are prices and quantities for three commodities in 1992 and 1993.

Commodity	1992		1993	
	Price	Quantity	Price	Quantity
Shirts	$22.00	6	$19.00	8
Shoes	$71.00	2	$78.00	3
Dresses	$53.00	8	$62.00	7

Compute the Laspeyres quantity index using 1992 as the base year.

Solution:

$$I_t = \frac{\sum q_{ti}\ p_{0i}}{\sum q_{0i}\ p_{0i}} \text{ x } 100$$

$$= \frac{8(22) + 3(71) + 7(53)}{6(22) + 2(71) + 8(53)} \text{ x } 100 = 108.88$$

Example 8 **Paasche Quantity Index**

Use the data in Example 7 to compute the Paasche quantity index. Use 1992 as the base year.

Solution:

$$I_t = \frac{\sum q_{ti}\ p_{ti}}{\sum q_{0i}\ p_{ti}} \text{ x } 100$$

$$= \frac{8(19) + 3(78) + 7(62)}{6(19) + 2(78) + 8(62)} \text{ x } 100 = 107.05$$

Example 9 **Fisher's Ideal Quantity Index**

Use your results from Examples 7 and 8 to compute Fisher's ideal quantity index.

Solution:

$$FI_t = \sqrt{\text{Laspeyres} \times \text{Paasche}}$$

$$= \sqrt{(108.88)(107.05)} = 107.96$$

Example 10 **Value Index**

Use the data in Example 7 to compute the value index. Use 1992 as the base year.

Solution:

$$I_t = \frac{\Sigma q_{ti} \; p_{ti}}{\Sigma q_{0i} \; p_{0i}} \times 100$$

$$= \frac{8(19) + 3(78) + 7(62)}{6(22) + 2(71) + 8(53)} \times 100 = 117.48$$

Example 11 **Stock Market Index**

Suppose the Little Stock Exchange (LYSE) trades only three stocks. Below are the prices for the stocks and the number of shares in 1992 and 1993.

	1992		1993	
Stock	Price	Shares	Price	Shares
Little Auto, Inc.	$27	47	$32	50
Little Phone Company	$18	100	$16	105
Little Computer	$92	72	$90	88

Calculate the market value weighted index and the price weighted index.

Solution:

$$\text{market- value weighted index} = \frac{\Sigma q_{ti}\ p_{ti}}{\Sigma q_{0i}\ p_{ti}} \times 100$$

$$= \left(\frac{32(50) + 16(105) + 90(88)}{27(47) + 18(100) + 92(72)}\right) \times 100 = 99.86$$

$$\text{Price- weighted index} = \frac{\Sigma P_{ti} / n}{\Sigma P_{0i} / n} \times 100$$

$$= \left(\frac{(32 + 16 + 90)/3}{(27 + 18 + 92)/3}\right) \times 100 = 100.73$$

Supplementary Exercises

Multiple Choice

1. If small stocks on the New York Stock Exchange (NYSE) rise more than larger stocks, then an equal weighted index of NYSE stocks would
   a. have a smaller return than the S&P 500 index.
   b. have a smaller return than a value weighted NYSE index.
   c. have a greater return than a value weighted NYSE index.
   d. have the same return as the value weighted NYSE index.
   e. cannot be determined by the information given.

2. A value weighted stock price index would be
   a. more sensitive to changes in small stocks than an equal weighted index.
   b. less sensitive to changes in small stocks than an equal weighted index.
   c. equally sensitive to changes in small stocks as an equal weighted index.
   d. equally sensitive to changes in large stocks as an equal weighted index.
   e. less sensitive to changes in large stocks than an equal weighted index.

3. Fisher's ideal price index
   a. is a Laspeyres index.
   b. is a Paasche index.
   c. is a simple average index.
   d. is a geometric average of a Laspeyres index and a Paasche index.
   e. always equals 1.

4. The Dow Jones Industrial Average is
   a. an index of consumer prices.
   b. a Laspeyres index.
   c. a Paasche index.
   d. an index of industrial production.
   e. a stock price index.

5. The base year
   a. is the reference year from which changes in the index are measured.
   b. is always the current year.
   c. is always last year.
   d. is the first year the index is created.
   e. is the last year the index is used.

6. A price relative is
   a. another name for the base year.
   b. another name for a Laspeyres index.
   c. another name for a Paasche index.
   d. the ratio of the price in the current year to the price in the base year.
   e. the ratio of the price in the base year to the price in the current year.

7. The GNP deflator is an example of a
   a. Laspeyres index.
   b. Paasche index.
   c. simple aggregate price index.
   d. base year price.
   e. current year price.

8. The FRB index of industrial production measures
   a. changes in the Dow Jones Industrial Average.
   b. changes in producer prices.
   c. changes in consumer prices.
   d. changes in the physical value of output.
   e. changes in consumer well being.

9. Which of the following is *not* one of the leading economic indicators?
   a. new orders for durable goods industries.
   b. corporate profits after taxes.
   c. index of stock prices for 500 common stocks.
   d. change in consumer prices.
   e. average hourly workweek.

10. The consumer price index is an example of a
    a. Laspeyres index.
    b. Paasche index.
    c. simple aggregate price index.
    d. base year price.
    e. current year price.

11. A large change in the price of commodities that are held in large quantities in the base year but are held in smaller quantities in the current year will lead to
    a. a greater change in the Paasche index relative to the Laspeyres index.
    b. a smaller change in the Paasche index relative to the Laspeyres index.
    c. an equal change in both the Paasche index and the Laspeyres index.
    d. a Laspeyres index equal to 0.
    e. a Paasche index equal to 0.

12. A large change in the price of commodities that are held in large quantities in the current year but are held in smaller quantities in the base year will lead to
    a. a greater change in the Paasche index relative to the Laspeyres index.
    b. a smaller change in the Paasche index relative to the Laspeyres index.
    c. an equal change in the Paasche index relative to the Laspeyres index.
    d. a Laspeyres index equal to 0.
    e. a Paasche index equal to 0.

True/False (if false, explain why)

1. An index number is a summary measure that allows for a comparison between a group of related items over time.
2. The most recent year is always used as the base year.
3. The consumer price index measures a market basket of goods.
4. A simple aggregate price index will not be affected by the units used to state prices.
5. A simple aggregate price index does not consider the relative importance of the commodities.
6. The Laspeyres index uses base year quantities.
7. The GNP deflator is an example of a Laspeyres index.
8. Fisher's ideal price index is a geometric average of the Laspeyres and Paasche indexes.
9. A quantity index measures the change in quantity from a base year to a particular year.

10. One problem with constructing a price index is determining the appropriate market basket.
11. The S&P 500 is an equal weighted stock market index.
12. A stock market index is a statistical measure that allows us to see how the prices of a group of stocks has changed over time.

Questions and Problems

Questions 1-10 use the following information:

Commodity	1993		1994	
	Price	Quantity	Price	Quantity
Beef	$2.50	50 lbs.	$2.75	40 lbs.
Chicken	$1.50	70 lbs.	$1.25	90 lbs.
Pork	$3.25	35 lbs.	$3.75	35 lbs.

Assume that 1993 is the base year.

1. Compute the price relatives for each good.
2. Construct the simple aggregate price index.
3. Construct the simple relative price index.
4. Construct the Laspeyres price index.
5. Construct the Paasche price index.
6. Use your results from Questions 4 and 5 to construct Fisher's ideal price index.
7. Construct the Laspeyres quantity index.
8. Construct the Paasche quantity index.
9. Use your results from Questions 7 and 8 to construct Fisher's ideal quantity index.
10. Construct the value index.

Answers to Supplementary Exercises

Multiple Choice

1. c	6. d	11. b
2. b	7. b	12. a
3. d	8. d	
4. e	9. d	
5. a	10. a	

True/False

1. True
2. False. Any year can be used as the base year, including the current year.
3. True
4. False. A simple aggregate price index will be affected by the units used to state the prices.
5. True
6. True
7. False. Paasche index.
8. True
9. True
10. True
11. False. Market value weighted index.
12. True

Questions and Problems

1. Price relative = P_{ti} / P_{0i}
   Beef = 2.75/2.50 = 1.100
   Chicken = 1.25/1.50 = .833
   Pork = 3.75/3.25 = 1.154

2. $$I_t = \frac{\sum p_{ti}}{\sum p_{0i}} \times 100$$
   $$= \frac{2.75 + 1.25 + 3.75}{2.50 + 1.50 + 3.25} \times 100 = 106.9$$

3. $$I_t = \frac{\sum p_{ti} / p_{0i}}{n} = \frac{1.10 + 0.833 + 1.154}{3} \times 100 = 102.9$$

4. $$I_t = \frac{\sum p_{ti}\ q_{0i}}{\sum p_{0i}\ q_{0i}} \times 100$$
   $$= \frac{2.75(50) + 1.25(70) + 3.75(35)}{2.50(50) + 1.50(70) + 3.25(35)} \times 100 = 103.64$$

5. $$I_t = \frac{\sum p_{ti}\ q_{ti}}{\sum p_{0i}\ q_{ti}} \times 100$$
   $$= \frac{2.75(40) + 1.25(90) + 3.75(35)}{2.50(40) + 1.50(90) + 3.25(35)} \times 100 = 101.43$$

6. $$F_t = \sqrt{\frac{\sum p_{ti}\ q_{ti}}{\sum p_{0t}\ q_{ti}} \ \frac{\sum p_{ti}\ q_{0i}}{\sum p_{0i}\ q_{0i}}} \times 100$$
   $$= \sqrt{(1.0614)(1.0143)} \times 100 = 103.76$$

7. $$I_t = \frac{\sum q_{ti}\ p_{0i}}{\sum q_{0i}\ p_{0i}} \times 100$$
   $$= \frac{40(2.50) + 90(1.50) + 35(3.25)}{50(2.50) + 70(1.50) + 35(3.25)} \times 100 = 101.45$$

8. $$I_t = \frac{\Sigma q_{ti}\ p_{ti}}{\Sigma q_{0i}\ p_{ti}} \times 100$$

$$= \frac{40(2.75) + 90(1.25) + 35(3.75)}{50(2.75) + 70(1.25) + 35(3.75)} \times 100 = 99.3$$

9. $$F_t \doteq \sqrt{\text{Laspeyres x Paasche}}$$

$$= \sqrt{(101.45)(99.3)} = 100.74$$

10. $$I_t = \frac{\Sigma q_{ti}\ p_{ti}}{\Sigma q_{0i}\ p_{0i}} \times 100$$

$$= \frac{40(2.75) + 90(1.25) + 35(3.75)}{50(2.50) + 70(1.50) + 35(3.25)} \times 100 = 102.91$$

CHAPTER 20

SAMPLING SURVEYS: METHODS AND APPLICATIONS

Chapter Intuition

Previously, we learned about the basic methods used in sampling. In many cases, simple random sampling may not represent the best method of selecting members from a population. This chapter continues the discussion of sampling by presenting methods that may be useful in designing a sampling survey. For example, when we are conducting a survey over a very large geographic region, it may be very expensive to conduct the survey. One way to deal with this problem might be to conduct a census over several smaller geographic areas. This approach is known as ***cluster sampling***. As another example, suppose we are interested in the earnings of people at a large company such as IBM. If we believe that the earnings of men and women in the company may be substantially different, we may want to ensure that both men and women are fairly represented in our sample. Because simple random sampling may not guarantee that both sexes are fairly represented, we may choose to use a method known as ***stratified sampling***. In stratified sampling, we divide the total population into different strata, in this case there will be two strata: men and women. A random sample of each group is then taken and analyzed. This approach ensures that each group in our population is fairly represented.

Chapter Review

1. From Chapter 8, we know sampling is preferred to a census for four reasons:
   a. sampling is more economical than a census.
   b. sampling is a much quicker way of obtaining data than a census.
   c. the large size of the population of interest may make a census infeasible.
   d. a census may be inappropriate for things like quality control.

2. The easiest approach to obtaining a sample is known as ***simple random sampling***. Simple random sampling is a technique in which each member of a population has an equal chance of being selected.

3. ***Stratified random sampling*** is a sampling technique in which the population is subdivided into groups known as ***strata***. This approach may be appropriate when the researcher believes that different members of the population will have different views. For example, a pollster may believe that the views on a civil rights bill may differ by racial and ethnic groups. In this case, the researcher may choose to use a stratified sampling technique in which the different strata are based on a person's racial or ethnic background.

4. When we sample, we incur different types of errors. ***Sampling errors*** occur when the difference between the sample and population parameters are due entirely to the items selected in the sample. ***Nonsampling errors*** are not connected to the type of sampling method used. They occur if the researcher chooses to sample the wrong population or if the responses are biased due to improperly worded questionnaires.

5. ***Cluster sampling*** is used when a researcher is interested in surveying a population which is spread over a large geographic region. In this case, the researcher may choose to divide the population into geographically compact clusters and then sample or take a census of a selected number of the clusters.

6. In ***two-phase sampling***, the researcher conducts a small pilot study in phase I before he or she attempts the larger scale study. The pilot study can be used to evaluate the questionnaire used and the number of nonrespondents.

7. Because the purpose of sampling is to make inferences about a population, we sometimes use ratios or regression analysis to improve our estimates.

8. Sometimes a technique known as the ***jackknife method*** is used in conjunction with sampling in order to remove the bias of an estimator and to produce confidence intervals.

Useful Formulas

Simple random sampling:

Sample mean:

$$\overline{X} = \frac{1}{n} \sum_{i=1}^{n} X_i$$

Sample variance:

$$s^2 = \frac{1}{n-1} \sum_{i=1}^{n} (X_i - \overline{X})^2$$

Population variance:

$$\sigma_{\overline{X}}^2 = \frac{s^2}{n} \times \frac{N-n}{N}$$

Confidence interval:

$$\overline{X} - z_{\alpha/2}\hat{\sigma}_{\overline{X}} < \mu < \overline{X} + z_{\alpha/2}\hat{\sigma}_{\overline{X}}$$

Simple random sampling for proportions:

Sample proportion:

$$\hat{p} = \frac{n}{N}$$

Estimate of population variance:

$$\hat{\sigma}_p^2 = \frac{\hat{p}(1-\hat{p})}{n-1} \times \frac{N-n}{N}$$

Confidence interval:

$$\hat{p} - z_{\alpha/2}\hat{\sigma}_p < p < \hat{p} + z_{\alpha/2}\hat{\sigma}_p$$

Stratified random sampling:

Sample mean for jth strata:

$$\overline{X}_j = \frac{1}{n_j}\sum_{i=1}^{n_j} X_{ij}$$

Sample size of proportional sampling:

$$n = \frac{\sum_{j=1}^{K} N_j \sigma_j^2}{N\sigma_{\overline{X}}^2 + \frac{1}{N}\sum_{j=1}^{K} N_j \sigma_j^2}$$

Estimate of population variance:

$$\hat{\sigma}_{\overline{X}}^2 = \sum_{j=1}^{n} W_j^2 \hat{\sigma}_{\overline{X}_j}^2$$

Proportion of the jth strata:

$$W_j = \frac{N_j}{N}$$

Sample variance for jth strata:

$$s_j^2 = \frac{1}{n_j - 1}\sum_{i=1}^{n_j} (X_{ij} - \overline{X}_j)^2$$

Confidence interval:

$$\overline{X} - z_{\alpha/2}\hat{\sigma}_{\overline{X}} < \mu < \overline{X} + z_{\alpha/2}\hat{\sigma}_{\overline{X}}$$

Optimal proportion for jth strata:

$$n_j = \frac{N_j \sigma_j}{\sum_{i=1}^{K} N_i \sigma_i} \times n$$

Optimal allocation for total sample:

$$n = \frac{\frac{1}{N}\left(\sum_{j=1}^{K} N_j \sigma_j\right)^2}{N\sigma_{\overline{X}}^2 + \frac{1}{N}\sum_{j=1}^{K} N_j \sigma_j^2}$$

Cluster sampling:

Sample mean:

$$\overline{X} = \frac{\sum_{i=1}^{m} n_i \overline{X}_i}{\sum_{i=1}^{n} n_i}$$

Estimate of variance:

$$\hat{\sigma}_{\overline{X}}^2 = \left[\frac{M - m}{M\, m\bar{n}^2}\right] \times \frac{\sum_{i=1}^{m} n_i^2 (X_i - \overline{X})^2}{m-1}$$

Confidence interval:

$$\overline{X} - z_{\alpha/2}\hat{\sigma}_{\overline{X}} < \mu < \overline{X} + z_{\alpha/2}\hat{\sigma}_{\overline{X}}$$

Ratio method:
$\hat{y}_r = \bar{y} / \bar{x}\, X$

Regression method:

$$\bar{y}_{lr} = \bar{y} + b\,(\overline{X} - \bar{x})$$

Example Problems

Example 1 Confidence Interval for the Population Mean

The Citizens for Fair Taxes of Rich City are interested in the average property tax paid by their 2,000 residents. A random sample of 25 of these households had a mean property tax of $3,222 with a standard deviation of $811.

a. Find an estimate of the variance of the sample mean.

b. Find a 90% confidence interval for the population.

Solution:

a. $$\hat{\sigma}_{\bar{X}}^2 = \frac{s^2}{n} \times \frac{N-n}{N}$$

$$\hat{\sigma}_{\bar{X}}^2 = \frac{811^2}{25} \times \frac{2000-25}{2000} = 25{,}979.98$$

b. $$\bar{x} \pm z_{.10/2}\hat{\sigma}_{\bar{X}}$$

$$3{,}222 \pm 1.645\sqrt{25{,}979.98}$$

$$2{,}956.85 \text{ to } 3{,}487.15$$

Example 2 Confidence Interval for the Population Mean

A fitness expert is interested in the mean number of miles marathoners run per week. Of the 100 members of the Crazy Legs Running Club, 40 were randomly sampled and found to run an average of 82.5 miles per week with a standard deviation of 12.1 miles. Find a 95% confidence interval for the mean number of miles that the runners run each week.

Solution:

$$\hat{\sigma}^2_{\overline{X}} = \frac{12.1^2}{40} \times \frac{100-40}{100} = 2.20$$

$$82.5 \pm 1.96\sqrt{2.20}$$

$$79.60 \text{ to } 85.40$$

Example 3 **Determining the Sample Size**

Suppose the quality control expert knows from past experience that the number of dud bullets in a case of 20,000 has a population standard deviation of 401. If she would like to compute a 95% confidence interval for the population mean with a standard deviation of 30, how many bullets should she sample?

Solution:

$$1.96\hat{\sigma}_{\overline{X}} = 30$$

$$\hat{\sigma}_{\overline{X}} = \frac{30}{1.96} = 15.31$$

$$n = \frac{N\sigma^2}{(N-1)\hat{\sigma}^2_{\overline{X}} + \sigma^2}$$

$$= \frac{2{,}000(401^2)}{(2{,}000-1)(15.31^2) + 401^2} = 508.52$$

So select a sample of 509.

Example 4 **Determining the Sample Size**

An auditor would like to estimate the total value of a corporation's accounts receivable. From previous years the auditor has found the population standard deviation to be $722 for the 1,200 accounts receivable. If the auditor would like to have a level of precision of $225, how large a sample should he select?

Solution:

$$n = \frac{1200(722^2)}{(1200-1)(225^2)+722^2} = 10.2$$

So select a sample of 11 (round up when selecting the sample).

Example 5 **Stratified Sampling-Proportional Allocation**

Suppose the auditor in Example 4 decides to divide the accounts receivable into stratum. If he would like a level of precision of $100, determine the total number of sample observations under a proportional allocation.

Stratum	Population Size	Standard deviation (estimated)
1	300	$85
2	375	$125
3	275	$50
4	250	$100

Solution:

$$N = 300 + 375 + 275 + 250 = 1200$$

$$n = \frac{\sum_{j=1}^{K} N_j \sigma_j^2}{N\sigma_{\bar{X}}^2 + \frac{1}{N}\sum_{j=1}^{K} N_j \sigma_j^2}$$

$$n = \frac{300(85^2) + 375(125^2) + 275(50^2) + 250(100^2)}{1200(10^2) + \frac{1}{1200}\left[300(85^2) + 375(125^2) + 275(50^2) + 250(100^2)\right]} = 86.70$$

So select a sample of 87.

Example 6 **Stratified Sampling-Optimal Allocation**

Use the data from Example 5 to determine the appropriate sample size under an optimal allocation rule.

Solution:

$$n = \frac{\frac{1}{N}\left(\sum_{j=1}^{K} N_j \sigma_j\right)^2}{N\sigma_{\bar{X}}^2 + \frac{1}{N}\sum_{j=1}^{K} N_j \sigma_j^2}$$

$$n = \frac{\frac{1}{1200}(300(85) + 375(125) + 275(50) + 250(100))^2}{1200(10^2) + \frac{1}{1200}\left[300(85^2) + 375(125^2) + 275(50^2) + 250(100^2)\right]} = 79.56$$

So select a sample of 80. Notice that selecting the optimal sample reduces the size of the sample over a proportional allocation.

Example 7 **Confidence Interval for Population Proportion**

Suppose the Fly Straight Golf Club Company sends its new golf clubs to 200 professionals. Of the 200 professionals receiving the clubs, 40 are randomly selected and asked if they hit the ball straighter with the new clubs. Twenty-two of the professionals said they did hit the ball straighter. Construct a 95% confidence interval for the population proportion.

Solution:

$$\hat{p} = \frac{n}{N}, \quad \hat{\sigma}_p^2 = \frac{\hat{p}(1-\hat{p})}{(n-1)} \times \frac{N-n}{N}$$

$$\hat{p} = \frac{22}{40} = .55, \quad \hat{\sigma}_p^2 = \frac{.55(1-.55)}{(40-1)} \times \frac{200-40}{200} = .0051$$

$$\hat{p} \pm z_{\alpha/2}\sigma_p^2$$

$$.55 \pm 1.96\sqrt{.0051}$$

$$.4103 \text{ to } .6900$$

Supplementary Exercises

Multiple Choice

1. In simple random sampling,
   a. all members of the population are examined.
   b. the population is divided into several groups and a census of all groups is taken.
   c. the population is divided into several groups and a census of some of the groups is taken.
   d. all members of the population have an equal chance of being selected.
   e. the population mean always equals the sample mean.

2. In cluster sampling,
   a. all members of the population are examined.
   b. the population is divided into several groups and a census of all groups is taken.
   c. the population is divided into several groups and a census of some of the groups is taken.
   d. all members of the population have an equal chance of being selected.
   e. the population mean always equals the sample mean.

3. In stratified random sampling,
   a. all members of the population are examined.
   b. the population is divided into several groups and a census of all groups is taken.
   c. the population is divided into several groups and a sample of some of all of the groups is taken.
   d. all members of the population have an equal chance of being selected.
   e. the population mean always equals the sample mean.

4. Sampling error is
   a. unimportant.
   b. the difference between the sample and population parameters due entirely to the sample.
   c. due to response error.
   d. due to nonresponses.
   e. due to measurement error.

5. Nonsampling error
   a. can result from misinformation provided by respondents.
   b. can result from nonresponses.
   c. is a type of measurement error.
   d. can result from choosing an inappropriate population to sample.
   e. all of the above.

6. Which of the following is *not* a step in survey sampling?
   a. determining the relevant information required for the study.
   b. constructing a population list in terms of the relevant population.
   c. collecting information on all population members.
   d. determining the method used to infer population parameters.
   e. drawing conclusions based on the sample information.

7. If we are surveying a population that is spread over a large geographic region, it is best to use
   a. a census.
   b. simple random sampling.
   c. stratified random sampling.
   d. cluster sampling.
   e. the jackknife method.

8. Suppose we are interested in studying the opinion of voters regarding a national day care policy. If we believe that people in different age groups will have different opinions, it is best to use
   a. a census.
   b. simple random sampling.
   c. stratified random sampling.
   d. cluster sampling.
   e. the jackknife method.

9. An unbiased estimate of the variance for the sample mean is
   a. s^2
   b. s^2/n
   c. $s^2/n\ [(N - n)/N]$
   d. $s^2/n\ [n/N]$
   e. $s^2/n\ [N/n]$

10. An unbiased estimator for the variance of the sample proportion is
    a. $[\hat{p}(1-\hat{p}) / n][(N-n) / N]$
    b. $[\hat{p}(1-\hat{p}) / (n-1)][(N-n) / N]$
    c. $[\hat{p}(1-\hat{p}) / (n-1)][N / (N-n)]$
    d. $[\hat{p}(1-\hat{p}) / (n-1)]$
    e. $\hat{p}(1-\hat{p})$

11. The sample mean for the overall population in stratified random sampling is
    a. $\bar{x}_j$
    b. $w_j\ x_j$
    c. $\Sigma\ w_j\ x_j$
    d. $\sum w_j \bar{x}_j$
    e. $\sum w_j \bar{x}_j / n$

True/False (if false, explain why)

1. For a given level of precision in stratified sampling, a proportional allocation will always require a larger sample than an optimal allocation.
2. The jackknife method is a method for removing bias in sampling.
3. The ratio method is used to remove bias in sampling.
4. A pilot study is the first stage in two-stage sampling.
5. Stratified random sampling is used when the population consists of several groups that are expected to have similar means.
6. Cluster sampling can reduce the costs of surveying a population that is spread over a large geographic region.
7. A random number table can be used to select a random sample.
8. A confidence interval can be constructed for the population mean, but not for the population total.
9. The ratio method can be used to forecast the population total.
10. Other things being equal, the smaller the level of precision desired, the smaller the sample size that needs to be collected.
11. A census will always be more accurate than a sample.
12. For a given population size and level of precision, the largest sample will need to be taken at a proportion of .5.

Questions and Problems

1. The Students for Affordable Education at Bargain College are interested in the average amount of money spent on textbooks by the college's 1,500 students. A random sample of 40 of these students had a mean of \$375 with a standard deviation of \$50. Construct a 95% confidence interval for the population mean.
2. Suppose a quality control expert knows from past experience that the number of defective widgets in a case of 50,000 has a population standard deviation of 725. If he would like to compute a 95% confidence interval for the population mean with a standard deviation of 50, how many widgets should he sample?

3. Suppose the Score High SAT Review Course is interested in the proportion of students that improve their scores after taking the course. Five hundred students have completed the course. Of the 500 students completing the course, 50 are randomly selected and asked if they improved their scores. Thirty-five of the students said they improved. Construct a 95% confidence interval for the population proportion.

4. The leader of People for Lower Taxes is interested in the proportion of citizens of Mayberry that favor the governor's tax reform policy. If there are 1,000 adults in Mayberry and the leader would like a 95% confidence interval to extend 5% on each side of the mean proportion, how many residents should be surveyed?

Answers to Supplementary Exercises

Multiple Choice

1. d
2. c
3. c
4. b
5. e
6. c
7. d
8. c
9. c
10. b
11. d

True/False

1. True
2. True
3. False. Used for forecasting.
4. True
5. False. Used when the means of the groups are expected to be different.
6. True
7. True
8. False. Confidence intervals can be constructed for the population mean and the population total.
9. True
10. False. The larger the sample size required.

11. False. Because some respondents may provide inaccurate information, a well designed sample survey may be more accurate than a census.
12. True

Questions and Problems

1. $\hat{\sigma}^2_{\bar{X}} = \frac{50^2}{40} \times \frac{1500-40}{1500} = 60.83$

$375 \pm 1.96\sqrt{60.83}$

359.71 to 390.28

2. $1.96\ \hat{\sigma}^2_{\bar{X}} = 50$

$\hat{\sigma}_{\bar{X}} = \frac{30}{1.96} = 25.51$

$$n = \frac{50{,}000\ (725^2)}{(50{,}000-1)\ (25.51^2) + 725^2} = 794.88$$

So select a sample of 794.

3. $\hat{p} = \frac{35}{50} = .70$

$\hat{\sigma}^2_{\hat{p}}\ \frac{.70(1-.70)}{(50-1)} \times \frac{500-50}{500} = .0038$

$.70 \pm 1.96\ \sqrt{.0038}$

.5792 to .8208

$1.96\ \sigma_{\hat{p}} = .05$

$\sigma_{\hat{p}} = .05 / 1.96 = .0255$

4.

$$n = \frac{Np\,(1-p)}{(N-1)\sigma_{\hat{p}}^2 + p(1-p)}$$

$$= \frac{1{,}000(.5)(1-.5)}{(1{,}000-1)(.0255^2)+.5(1-.5)} = 277.9$$

So select a sample of 278.

CHAPTER 21

STATISTICAL DECISION THEORY: METHODS AND APPLICATIONS

Chapter Intuition

Throughout the book, we have learned about the role of statistics in dealing with an uncertain world. The ultimate goal of using statistics is to allow us to make better decisions when the future is uncertain. For example, suppose you are wrongly accused of committing a crime in a different country. The penalties for the crime are a $25 fine if you plead guilty, or 5 years in jail if you go to trial and lose. Although most of us would not like to plead guilty to a crime we did not commit, the uncertainty of the trial system in another country and the severe penalty if we should fight and lose, would lead most of us to pay the fine and leave the country as fast as we could. In this case, we have used a decision making approach known as the ***minimax criterion***. In the minimax criterion, we assume that whatever action we take, the worst possible outcome will prevail. We then try to minimize the worst thing that can happen to us (maximum penalty). In the crime example, by pleading guilty, we have limited the maximum penalty to a $25 fine. As another example, suppose an owner of a small business is trying to decide whether to lease a large or small machine. Although leasing a large machine can generate higher profits when future demand is high, it can lead to larger losses when future demand is low. The owner can address this problem in two ways. First, he can use an approach that ignores the probabilities of future states of the world, such as the minimax, ***maximax***, and ***maximin*** methods. Second, he can incorporate the probability of future states of the world into decision making. The ***expected monetary value***, ***utility criterion***, and ***Bayesian method*** are approaches that incorporate probabilities into the decision making. This chapter deals with different strategies for making decisions under uncertainty.

Chapter Review

1. Some of the decisions facing a business manager are:
   a. What products should be produced?
   b. What investments should be purchased?
   c. What projects should be undertaken by the firm?

2. The four key elements in the decision-making process are:

   a. ***Action***: the choices available. For example, should I purchase insurance or not?

   b. ***State of nature***: the uncertain elements of a decision. For example, will I get into an accident or not?

   c. ***Outcomes***: the consequences of each combination of an action and a state of nature. For example, I buy insurance and I have an accident or I buy insurance and I don't have an accident.

   d. ***Probability***: the chance that a state of nature occurs. For example, there is a 1% chance that I will get into an accident.

3. There are two methods of decision making that do not use probabilities: The ***maximin*** and the ***minimax regret criterion***. The maximin criterion is a very conservative strategy. It assumes that the worst outcome will occur regardless of what action is taken. Using the maximin criterion, we would choose the action that results in the best outcome, given that the worst state will occur. To use the minimax criterion, we have to generate the ***regret matrix***. We then consider that the worst outcome (maximum regret) will occur for whichever action is taken. The decision then is to choose the action that has the smallest maximized regret.

4. Decision making can also use probabilities. For each action, the payoffs are different depending on the state of nature that occurs. One way to evaluate the actions is to calculate the expected payoffs of each action. The criterion is called the ***expected monetary value criterion*** (EMV).

5. The probability used to calculate the EMV can be improved by incorporating sample information. The original probability distribution before the revision is called the prior distribution. The revised probability is called the posterior distribution.

6. Very often, it is the satisfaction (in economic terms ***utility***) from the monetary value rather than the monetary value itself that determines how people make decisions. Decision making based on satisfaction or utility is called the ***expected utility criterion***. The expected utility is found by averaging the amount of satisfaction a person receives over the different possible states of nature.

7. One simple approach for making decisions is the use of ***decision trees***. The decision tree is similar to the ***probability tree*** that was previously discussed. The advantage of using a decision tree is that all possible outcomes can be easily seen at once.

Useful Formulas

Bayes' theorem:

$$P_r(E_2 | E_1) = \frac{P_r(E_1 | E_2)\, P_r(E_2)}{P_r(E_1)}$$

Capital market line:

$$E(R_p) = R_f + [E(R_m) - R_f]\frac{\sigma_p}{\sigma_m}$$

Capital asset pricing model:

$$E(R_i) = R_f + \beta_i [E(R_m) - R_f]$$

Systematic risk:

$$\beta_i = \frac{\mathrm{Cov}(R_i, R_m)}{\mathrm{Var}(R_m)} = \frac{\rho_{i,m}\,\sigma_i\,\sigma_m}{\sigma_m^2} = \frac{\rho_{i,m}\,\sigma_i}{\sigma_m}$$

Example Problems

Example 1 **Maximin Criterion**

A copy center is considering whether it should lease a large machine or a smaller machine for the next year. The net profits for leasing these two machines are reported in the following table.

	Action	
State of nature	Lease a large machine	Lease a small machine
High demand	30	25
Medium demand	10	10
Low demand	–5	5

Use the maximin method to determine the best action.

Solution:

Step 1: Find the minimum payoff for each action. The minimum payoff occurs when we have low demand for both actions.

Step 2: Select the action that has the maximum payoff from step 1. In this case we choose the small machine because the payoff of 5 is greater than the –5 payoff for the larger machine.

Example 2 **Minimax Regret Criterion**

Use the data given in Example 1 to find the best action based on the minimax regret criterion.

Solution:

Step 1: Obtain the regret matrix, which is the optimal payoff minus the actual payoff for each state of nature.

	Action	
State of nature	Large machine	Small machine
High demand	0	5
Medium demand	0	0
Low demand	10	0

Step 2: Find the maximum regret (highest values) for each action.

Step 3: Choose the action that gives the smaller maximum regret. In this case, we choose to the small machine.

Example 3 **Minimax Strategy**

A homeowner is considering whether she should insure her house completely, partially, or not at all. The loss matrix is:

State of nature	Fire	No fire
Not insure	100,000	0
Partially insure	51,000	1,000
Fully insure	2,000	2,000

What is the best action under a minimax strategy?

Solution: Notice that in this example, the number is the cost, not the payoff. Therefore, a minimax strategy is a conservative strategy.

Step 1: Find the maximum loss for each action. The maximum losses for not insured and partially insured would be the fire state of nature. For the no fire state, both would have the same loss.

Step 2: Choose the action that minimizes the maximum loss. In this case, the choice is to fully insure.

> Example 4 **Maximin Strategy**
>
> Use the information given in Example 3 to find the best action using a maximin criterion.

Solution:

Step 1: Construct the regret matrix.

98,000	0
49,000	1,000
0	2,000

Step 2: Maximize the regret for each action.

Step 3: Choose the action that minimizes the maximum regret. In this case, the choice is to fully insure.

Example 5 **Expected Monetary Value**

A company is considering the following three options for an assembly line: i) do nothing; ii) upgrade the old assembly line; and iii) install a new one. The payoff matrix under different demand situations is presented in the following table:

State of nature	Do nothing	Upgrade	Install new	P_r
Low demand	10	5	0	0.3
Moderate demand	10	15	15	0.3
High demand	10	25	30	0.4

Compute the expected monetary value of each action.

Solution:

E(do nothing) = 10(.3) + 10(.3) + 10(.4) = 10

E(upgrade) = 5(.3) + 15(.3) + 25(.4) = 16

E(install new) = 0(.3) + 15(.3) + 30(.4) = 16.5

The best action is to install the new machine because it has the highest EMV.

Example 6 **Expected Monetary Value**

An ice cream vendor at the local beach knows from past experience that the sales of ice cream on a Sunday depend on the weather. He also knows from past experience that the probability of sunny weather is 0.8 and the probability of rainy weather is 0.2. The net profit from ordering 50 pounds and 100 pounds of ice cream are presented in the following table.

Weather	(A1) 50 lbs.	(A2) 100 lbs.	Probability
(S1) Sunny	100	150	0.8
(S2) Rainy	80	50	0.2

Compute the expected monetary value for each action.

Solution:

$$EMV_{A1} = 100(.8) + 80(.2) = 96$$

$$EMV_{A2} = 150(.8) + 50(.2) = 130$$

Example 7 **Bayesian Method**

Now suppose that the ice cream vendor of Example 6 uses the weather report on Saturday evening. The weather report on Saturday is for rain on Sunday. The weatherman's track record is summarized below.

$P_r(I1 \mid S1) = 0.8$ $P_r(I2 \mid S1) = 0.2$

$P_r(I1 \mid S2) = 0.3$ $P_r(I2 \mid S2) = 0.7$

where I1 indicates a sunny report.
I2 indicates a rainy report.

Use this information to obtain the posterior probability. What is the expected monetary value using the posterior distribution?

Solution:

$$P_r = \frac{P_r(S1 \cap I1)}{P_r(I1)} = \frac{P_r(I1 \mid S1)\ P_r(S1)}{P_r(I1 \mid S1)\ P_r(S1) + P_r(I1 \mid S2)\ P_r(S2)}$$

$$= \frac{0.8\ (0.8)}{0.8(0.8) + 0.3(0.2)} = 0.914$$

$P_r(S2 \mid I1) = 0.086$

$$\frac{P_r(S1 \cap I2)}{P_r(I2)} = \frac{P_r(I2 \mid S1)\ P_r(S2)}{P_r(I2 \mid S1)P_r(S1) + P_r(I2 \mid S2)P_r(S2)}$$

$$= \frac{0.2(0.8)}{0.2(0.8) + 0.7(0.2)} = 0.533$$

$P_r(S2 \mid I2) = 0.467$

The expected monetary value given I2 is
$E(A1 \mid I2) = 100(.533) + 80(.467) = 90.69$

E(A2 | I2) = 150(.533) + 50(.467) = 103.3

The expected monetary value given I1 is

E(A1 | I1) = 100(.914) + 80(.086) = 98.28

E(A2 | I1) = 150(.914) + 50(.086) = 141.4

Example 8 **Utility Criterion**

A farmer harvests 1,000 bushels of wheat every year. The total revenue from wheat is 1,000P, where P is the price of wheat at the time of the harvest. His utility function is

$$U = (1{,}000P)^{1/2}$$

Suppose the farmer knows that there is a 50% probability that the price will be 100 and a 50% probability that the price will be 80. The farmer decides to sell his harvest using the futures market and is guaranteed a price of 90.

Is this behavior consistent with his utility function? Is the farmer risk averse, risk neutral, or a risk taker?

Solution: The utility of selling in the futures market is

$$U = [1{,}000P]^{1/2} = [1{,}000(90)]^{1/2} = 300$$

The expected utility of not selling in the futures market is

$$[1{,}000(100)]^{1/2} \times .50 + [1{,}000(80)]^{1/2} \times .50 = 299.54$$

Because 300 > 299.54, his choice is consistent with his utility function. Because

E(U) < U[E(1,000P)], he is risk averse.

Example 9 **Expected Utility**

Consider the following three projects.

Project A		Project B		Project C	
Payoff	P_r	Payoff	P_r	Payoff	P_r
5	1/3	4	1/3	5	1/4
10	1/3	10	1/3	10	1/2
15	1/3	16	1/3	15	1/4

Which is preferred if $U(W) = W - 1/20\ W^2$?

Solution: Use the expected utility to solve this problem. The expected utility for each project is

$$E[U(W)] = \Sigma \Pr_i U_i = 1/3[5 - 1/20\ 5^2] + 1/3[10 - 1/20\ 10^2] + 1/3[15 - 1/20\ 15^2] = 4.17$$

$$E[U(W)] = \Sigma \Pr_i U_i = 1/3[4 - 1/20\ 4^2] + 1/3[10 - 1/20\ 10^2] + 1/3[16 - 1/20\ 16^2] = 3.8$$

$$E[U(W)] = \Sigma \Pr_i U_i = 1/4[5 - 1/20\ 5^2] + 1/2[10 - 1/20\ 10^2] + 1/4[15 - 1/20\ 15^2] = 4.17$$

Even though the expected monetary values are the same for each project, the expected utilities are different. The expected utilities are different because of differences in risk (as measured by the standard deviation). Project C has the highest expected utility because it has the smallest risk.

Supplementary Exercises

Multiple Choice

1. The minimax strategy
   a. uses probabilities.
   b. uses Bayes' Theorem.
   c. maximizes the worst thing that can happen.
   d. minimizes the maximum regret.
   e. is another name for a decision tree.

2. The EMV criterion
   a. maximizes the satisfaction of the decision maker.
   b. uses Bayes' Theorem.
   c. maximizes the worst thing that can happen.
   d. minimizes the maximum regret.
   e. looks at the expected value of the payoffs.

3. Which of the following is *not* an element of the decision making process?
   a. action.
   b. state of nature.
   c. outcome.
   d. probability.
   e. CAPM.

4. The expected utility criterion is computed by
   a. $\Sigma\, Pr_i$
   b. $\Sigma\, U_i$
   c. $\Sigma\, Pr_i\, U_i$
   d. $\Sigma\, Pr_i / U_i$
   e. $\Sigma\, U_i / Pr_i$

5. The capital asset pricing model is a decision-making model in finance that
   a. shows the tradeoff between risk and expected return for combinations of a riskless asset and the market portfolio.
   b. shows the relationship between the expected return on a stock and its total risk.
   c. shows the expected price of an asset based on its total risk.
   d. shows the relationship between the expected return on an asset and its nonsystematic risk.
   e. shows the relationship between the expected price of an asset and its nonsystematic risk.

6. Utility refers to
   a. the amount of satisfaction a person receives.
   b. the amount of money a person receives.
   c. the expected amount of money a person receives.
   d. the minimax strategy.
   e. the maximin strategy.

7. Which of the following statements is true?
   a. The posterior distribution contains less information than the prior distribution.
   b. The posterior distribution is derived from the prior distribution and the sample information.
   c. The prior information is derived from the posterior distribution and the sample information.
   d. The sample information is derived from the prior distribution.
   e. The sample information is derived from the posterior distribution.

8. If $P_r(S_1) = P_r(S_2)$, then
   a. $P_r(S_1 \mid I_1) = P_r(S_1 \mid I_2)$
   b. $P_r(S_1 \mid I_2) = P_r(S_1 \mid I_1)$
   c. $P_r(S_2 \mid I_1) = P_r(S_1 \mid I_2)$
   d. $P_r(S_1 \mid I_2) = P_r(S_2 \mid I_2)$
   e. the posterior distribution equals sample information.

9. A risk averter
   a. never takes risk.
   b. takes risk with constant increase of expected value.
   c. takes risk with increasing increase of expected value.
   d. takes risk with decreasing increase of expected value.
   e. takes risk for no good reason.

10. A risk lover
   a. never takes risk.
   b. takes risk with constant increase of expected value.
   c. takes risk with increasing increase of expected value.
   d. takes risk with decreasing increase of expected value.
   e. takes risk for no good reason.

True/False (if false, explain why)

1. The EMV criterion for decision making does not use probability.
2. The EMV criterion looks at the expected value of the money received to determine the best action to take.
3. The minimax regret strategy is a conservative approach to decision making.
4. Decision trees can be used in decision making.
5. The capital market line is used to show the trade-off between risk and return for combinations of the market portfolio and the risk free asset.
6. The maximin and minimax approaches to decision making both use probabilities.
7. The utility approach to decision making looks at the expected value of the money received in order to determine the best action to take.
8. A risk averter will never take risk.
9. For a risk averter, the expected utility of wealth is lower than the utility of expected wealth.
10. The maximin strategy is an aggressive strategy.
11. The Bayesian method of decision making uses sample information to revise the prior distribution into posterior distribution.
12. The expected monetary value approach is equivalent to the expected utility approach, when an individual is risk neutral.
13. A risk averter prefers lower risk to higher risk given the same level of expected value.
14. A risk lover prefers higher risk (standard deviation) than lower risk given the same level of expected value.

Questions and Problems

1. Below is the payoff matrix for a delivery company trying to decide if it should buy a new delivery truck or repair the old truck.

State	New Truck	Repair Old	Prob
Few miles driven	45	100	1/3
Average miles driven	70	70	1/3
Many miles driven	125	80	1/3

   Compute the expected monetary value for each action.

2. Suppose the owner of the delivery company in Problem 1 has a utility function of $U = W - W^{1/2}$. Find the expected utility for each action.

3. Bob Jones is in the market for a new car. He knows that he will pay one of two possible prices, \$10,000 or \$12,000. Suppose Bob's utility function is $U = P^{-.1}$. If Bob believes there is a 20% probability of getting the \$10,000 price and an 80% probability of getting the \$12,000 price, compute Bob's expected utility.

4. Suppose a company is trying to decide whether to replace an old machine with a new machine. Because the new machine is more efficient, it will result in higher profits if sales are high. However, if sales are low the cost of the machine will result in losses to the company. Below is a table of the profits for the two possible actions.

	Action	
State of demand	New machine	Old machine
Low	–50	50
Average	125	100
High	175	60

   Use the maximin criterion to decide on the best course of action.

5. Reconsider Problem 4. Use the minimax regret criterion to decide on the best course of action.

Answers to Supplementary Exercises

Multiple Choice

1.	d	6.	a
2.	e	7.	b
3.	e	8.	e
4.	c	9.	c
5.	d	10.	d

True/False

1. False. Does use probabilities.
2. True
3. True
4. True
5. True
6. False. Does not use probabilities.
7. False. Uses the expected value of utility.
8. False. A risk averter will take risk as long as the gain from taking risk outweighs the loss from taking the risk.
9. True
10. False. In doing maximin, we assume that the worst situation will occur in the first step; then we pick the best out of all worst situations. So, it is a very conservative strategy.
11. True
12. True
13. True
14. False. A risk lover has to be compensated with decreasing expected value for an additional unit of risk taken.

Questions and Problems

1. EMV for the new truck is
   $$EMV = 1/3(45) + 1/3(70) + 1/3(125) = 80$$
   EMV for repairing the old truck is
   $$EMV = 1/3(100) + 1/3(70) + 1/3(80) = 83.33$$
   Repair the old truck using the EMV criterion.

2. For the new truck
   $$E(U) = 1/3[45 - 45^{1/2}] + 1/3[70 - 70^{1/2}] + 1/3[125 - 125^{1/2}] = 71.25$$
   For repairing the old truck
   $$E(U) = 1/3[100 - 100^{1/2}] + 1/3[70 - 70^{1/2}] + 1/3[80 - 80^{1/2}] = 56.99$$
   Select the new truck based on the expected utility criterion.

3. The expected utility is
   $$E(U) = .20(10{,}000^{-.1}) + .80(12{,}000^{-.1}) = .3923$$

4. The minimum profit occurs in the low demand state for both courses of action. The minimax criterion will then tell us to select the action that has the maximum profit from the low demand state. In this case, we select keeping the old machine.

5. To use the maximin regret criterion, we need to obtain the regret matrix, which is found by taking the optimal payoff minus the actual payoff in each state.

Regret

State of demand	New machine	Old machine
Low	100	0
Average	0	25
High	0	115

The maximum regrets in the regret matrix are for the new machine in the low demand state and for the old machine in the high demand state. The maximin regret criterion tells us to select the action that has the minimum maximum regret. In this case, we should choose the new machine.